I0486409

Germs 101

A Mini-Course for Everyone

By

Darralyn McCall

This book is a work of non-fiction. Names and places have been changed to protect the privacy of all individuals. The events and situations are true.

© 2003 by Darralyn McCall. All rights reserved.

No part of this book may be reproduced, stored in a retrieval system, or transmitted by any means, electronic, mechanical, photocopying, recording, or otherwise, without written permission from the author.

ISBN: 1-4107-2087-X (e-book)
ISBN: 1-4107-2088-8 (Paperback)

Library of Congress Control Number: 2003090949

This book is printed on acid free paper.

Printed in the United States of America
Bloomington, IN

1stBooks - rev. 05/07/03

Dedication

This is dedicated to: my sister Andrea and brother-in-law Gary, the Boxer kids and my grandparents. This is also dedicated to some special teachers, Dan Caldwell, Jean Cooper and John Baranway. Finally, this is dedicated to Laurie Krumrey, a special friend who asked the important questions.

Table of Contents

Introduction

This is a "mini-course" in microbiology for the non-biologist. The purpose of this book is to get everyone on the "same page" in their understanding of germs and what they do. There is a lot of information being tossed at you on a daily basis and some of it is wrong. That wrong information is passed around because someone didn't do his homework and get all the facts.

There are informative books on microbiology and biological terrorism, such as the book by Senator Bill Frist, M.D.. You need a basic knowledge of microbiology to get the complete story from those books. Read the other books, but read this book first.

From reading this book, you'll learn:

- some general biology
- the different types of germs and the names of the scientists who study them
- how germs spread
- how your body fights germs
- how germs invade your body
- how to kill germs
- how to handle raw food and other contaminated material safely

- why staying home when you have a cold is good for everyone
- the names of infectious diseases and their symptoms
- some background information on the germs that are in the news
- where to find more information

This book doesn't cover how to do any of the laboratory work involved in microbiology, such as how to grow germs, because unless you plan to become a lab technician, you don't need to know that. This book also doesn't cover the material you'd need to pass a college course in microbiology, because that isn't the intent of this book.

This book requires no previous training in biology. If you have a background in biology, you'll probably be bored by this book. If you have a background in biology, you'll also notice that I take some liberties in the interest of making the concepts of microbiology easier for a non-biologist to understand.

There is no information on SARS, because the story is still unfolding. You can get the most up-to-date information on SARS by going to the CDC's web site at: http://www.cdc.gov. Currently, the best way to protect yourself against SARS is to practice every day prevention, covered in Chapter three of this book.

Chapter One
Introduction to Biology

Cells—the basic unit of life

Most biology courses and textbooks will tell you that a **cell** is the basic unit of life. Guess what? That's exactly what I'm going to tell you.

Cells can be roughly divided into two categories, **eukaryotic** and **prokaryotic**. Plants, animals (including humans), algae, parasites and fungi are examples of eukaryotic cells. Bacteria are prokaryotic cells.

Viruses aren't cells at all, because they don't have the structures found in cells and they aren't able to do "cell activities." **Viruses** are often referred to as **particles**. "Cell activities" include eating and drinking which are termed phagocytosis and pinocytosis when a cell does it, eliminating waste products, reproduction, rebuilding cell structures and other things.

People, as well as many plants and animals are made up of millions of cells. Many of the cells are assembled like brick walls, into tissues, such as skin, muscle, organs. In tissue, the cells are the bricks and a substance called **hyaluronic acid** acts like the mortar that holds the bricks together. You'll learn more about hyaluronic acid in the chapter on Enzymes and Toxins.

1

There are some animals and plants that consist of only 1 cell. Bacteria are single cells. Some types of bacteria grow together so they may look like they are more than one cell.

What are cells made of?

What are cells made of? All cells have a **cell membrane**. Another name for cell membrane is **cytoplasmic membrane**, because it contains the **cytoplasm** that all cells have. The terms cytoplasmic membrane and cell membrane can be used interchangeably.

Cell membranes are mostly fats, or if you want to use the scientific name, **lipids**. There are some islands of protein floating around in the fats of the cell membranes. Cell membranes are described as **semipermeable**, because they allow some substances to flow through them, but keep other substances on one side or the other of the membrane. Water flows through the membranes easily.

Cytoplasm is the stuff inside a cell. The membrane keeps the cytoplasm inside the cell. You may have heard that people are 90+% water and that's because the cytoplasm is mostly water. You'll also find vitamins, minerals, protein, fats, carbohydrates and other chemicals in the cytoplasm.

You can make a model of a cell membrane and the cytoplasm at home or in your classroom. Take a plastic bag to represent the membrane. Fill the plastic bag with water, toss in some sugar, a multiple vitamin and mineral tablet, some salt, some butter or vegetable oil and some protein powder and you have a model of cell membrane and the cytoplasm.

Plant cells and most bacteria have **cell walls** outside the cell membrane. Animal and human cells don't have cell walls. You can represent the cell wall by putting your model of the cell membrane and cytoplasm into a box.

Cells also have structures inside. These structures generate energy for the cell, repair the cell, and allow the cell to divide and duplicate itself. The structures are, for the cell, what your organs are for your body. This is why the structures inside the cell are called **organelles**.

The organelles inside plant and animal cells are more complex than the organelles inside bacteria. There are more different types organelles inside plant and animal cells than in bacterial cells. The organelles inside plant and animal cells include mitochondria, cristae, endoplasmic reticulum, the Golgi apparatus and ribosomes. Bacteria have ribosomes, but not the mitochondria, cristae, endoplasmic reticulum or Golgi apparatus. The ribosomes in bacterial cells are different from the ribosomes in plant and animal cells.

All cells have **DNA, Deoxyribose Nucleic Acid**. **Chromosomes** are DNA. The DNA in eukaryotic cells is located inside another membrane, called the **nuclear membrane**. The area inside the nuclear membrane is called the cell's **nucleus**. The DNA in bacteria is just floating around in the cytoplasm, though it often hangs out by the cell membrane.

All cells have **RNA**, **Ribose Nucleic Acid**. The cells use the RNA to **translate** the DNA code into proteins. To get an idea of how this works, think of the ribosomes as the heads of a video tape player and the RNA as the tape. The protein that comes from the ribosomes reading the RNA, is like the image of the movie on the television screen that you get when the tape player's heads read the videotape.

Cell proteins may be **structural**, and support parts of the cell like steel girders support a building, or they may be **enzymes**. Some enzymes digest food. Other enzymes determine the characteristics of the structures or the organism. All enzymes help chemical reactions to happen. An example of an enzyme that digests food, is a protease. A protease is an enzyme that digests protein. An example of an enzyme that determines a characteristic is the enzyme that causes a person's eyes to be blue or hair to be brown.

Cells generate energy and use energy. The main energy storage chemical in a cell is **Adenosine Triphosphate**, better known as **ATP**. As you have probably guessed from its name, ATP has three phosphates attached to it. When ATP is converted to **ADP**, **Adenosine Diphosphate** (2 phosphates), energy is released from storage so the cell can use it. ATP is like a rechargeable battery that has been charged. The charging process is where phosphates are attached to ADP to create ATP. The battery discharges as ATP molecules are converted to ADP and separate phosphate molecules.

Cells of bacteria, fungi, algae, parasites, plants and animals use ATP to store energy. Viruses don't have the mechanisms for using or storing ATP.

Some nutrients provide the building blocks for the cell structures and are "burned" for energy. Other nutrients, such as vitamins are attached to the protein part of an enzyme to make it active. Minerals like calcium, magnesium and phosphorus (often in the form of phosphate) serve other purposes in the cells.

Humans and animals require oxygen. Some germs need oxygen as well. Cells that require oxygen create carbon dioxide as a waste product. Some germs use other chemicals instead of oxygen. Sulfur is a common substitute for oxygen in bacteria. The germs that use sulfur, produce hydrogen sulfide as a waste product instead of carbon dioxide.

The germs that use oxygen are called **aerobic organisms**. The germs that don't use oxygen are called **anaerobic organisms**.

As stated above, **viruses aren't cells**. They don't have cell membranes, cytoplasm or the structures that generate energy or allow them to reproduce themselves. Viruses hijack a host cell and force it to make more viruses. Viruses only have a few enzymes—the ones that allow them to take over their host cell and force it to make more viruses. Viruses only have **DNA *or* RNA**. They never have both.

The differences between germ cells and human or animal cells are important. The differences are a key factor in selecting chemicals for use as antibiotics. You'll learn more about this in the chapter on antibiotics.

Is this all there is to know about cells? Not even close. Entire textbooks are devoted to cells. This is enough information about cells for the purposes of this book.

Chapter Two
Of Scientists and Germs

You can't turn on the television or flip through a magazine or newspaper without encountering something about **germs**. Naturally occurring germs and the threat of germ or biological warfare are in the news every day in reports about anthrax, *E. coli,* mad cow disease and smallpox virus.

Most of the household cleaning products emphasize the fact that they "kill germs" in their advertising. You'll also see advertising about cold sore medication or medication for athlete's foot or toenail fungus or vaginal yeast infection medication.

Escherichia coli, better known as *E. coli*, is a germ. Anthrax is caused by another germ, *Bacillus anthracis.* Yeast infections are caused by *Candida albicans*, yet another germ. Athlete's foot fungus and toenail fungus are also germs. Cold, flu and cold sore viruses are germs. Mad cow disease and hoof and mouth disease are caused by germs.

When you hear the word, germs, you probably think of horrible little creatures that cause diseases. Disease producing germs may be **bacteria**, **viruses**, **fungi**, **algae** or **parasites** and are grouped into the category of **pathogenic** organisms. Not all bacteria, algae or fungi cause diseases. The bacteria, algae and fungi that *don't* cause disease are categorized as

7

non-pathogenic organisms. There are some organisms that usually don't cause disease, but they can. These organisms are categorized as **opportunist pathogens** because they cause disease when they are given the *opportunity* to do so.

Let's get acquainted with the various types of germs and the scientists who study them. All living things can be called **organisms**. Bacteria are organisms, dogs are organisms, you and I are organisms. The general term for germs and other organisms that are too small to be seen without a microscope is **microorganism** or **microbe**. The general term for the scientist who studies microorganisms is: **microbiologist**.

Bacteria and Bacteriologists:

The first type of germ is the **bacterium**. The plural of bacterium is **bacteria**. The people who study them are either microbiologists, or specifically, **bacteriologists**.

Note to writers: bacteria have two parts to their names, *Escherichia coli* for example. *Escherichia* is the **genus name**— sort of like the germ's "last name" and *coli* is the **species name** or "first name." Note that the Genus name starts with a capital letter and the species name doesn't. The genus name may be shown as the first letter of the genus name, followed by a period, such as *"E."* for *Escherichia*. If there is any chance of confusion, it's best to write out the entire genus name.

You've probably also noticed that the genus and species name are shown in italics here. If you write about a bacterium, the correct way to show the name is either in italics or underlined, such as <u>Escherichia</u> <u>coli</u>. When the "whole bacterium family" is referenced, you may write *Bacillus species*, referring to the clan *Bacillus* without identifying individual members.

Bacteria are 1 celled organisms that can be seen under a standard (aka light) microscope. This is the kind of microscope used in high school biology and is available at the local hobby store.

Bacterial cells and human cells need different nutrients. Folic acid, a B-vitamin, is an example of a nutrient required by human cells. Many bacteria make folic acid. Some of the bacteria that make folic acid live in your intestines. You share the food you eat with them and they share the folic acid they make with you. You may have guessed from this information that these bacteria are non-pathogenic or at worst, opportunist pathogens.

Some bacteria produce **spores**. Human cells do not produce spores. Spores are a resting state that some bacteria can go into. This is sort of like **hibernation**. Another analogy for spores is plant **seed**. Spores don't need the food and water that bacteria in a non-spore state need.

You've heard a lot about bacterial spores from the news. Some bacteria form spores that can survive in conditions that the vegetative bacterial cell can't survive. **Vegetative cells** are the bacterial cells that are actively eating and drinking and doing all the normal cell activities. Spores are hibernating, waiting for conditions that provide the food, water and other conditions they need to return to the vegetative state, like bears hibernating during the winter when their food is scarce.

The ability to produce spores gives a **survival advantage**. The bacteria that produce spores can survive conditions that non-spore-forming bacteria can't survive.

That survival advantage is why you've heard news reports suggesting that the **anthrax spores** sent in the letters in fall 2001, could be decades old, though more recent reports give the age as around 2 years old for those spores.

Once the anthrax spores got into the victims' bodies, they found the appropriate conditions to return to the vegetative state, like a seed germinating when it's planted. The cutaneous and pulmonary anthrax resulted from the spores returning to an active vegetative state and reproducing themselves. The vegetative cells produced **toxins** that caused many of the symptoms of anthrax. The spores were not able to produce the toxins.

Some of the bacteria that produce spores are the genus *Bacillus*, whose best-known member is *Bacillus anthracis* and causes anthrax. Other members of the genus *Bacillus* include *Bacillus subtilis var niger* and *Bacillus stearothermophilus*. These are organisms that rarely cause disease.

Bacillus subtilis var niger is used as an **"indicator" organism** to test the efficiency of ethylene oxide sterilizers in hospitals and medical device and pharmaceutical manufacturing facilities, because its spores are extremely hard to kill. When used as the indicator organism, killing the spores of *Bacillus subtilis var niger* **indicates** that the sterilization process is adequate to kill off any contaminating bacteria. *Bacillus stearothermophilus* is used as an indicator organism to test the efficiency of steam autoclaves. It indicates that the sterilization process is adequate to sterilize surgical equipment and intravenous solutions, such as the 'D5W' (5% dextrose in water) or saline that you may have heard someone request on the television medical shows.

Another genus of bacterium that produces spores is *Clostridium*. The two best-known members of this genus are: *Clostridium botulinum* and *Clostridium tetani*. You can probably tell from their names that the first member of this genus causes botulism and the second one causes tetanus. Other members of the genus *Clostridium* cause disease as well, but aren't as well known. *Clostridium perfringens* causes gas-gangrene if it infects your body. It causes food poisoning if it grows in the stuffing that is cooked inside the Thanksgiving turkey. *Clostridium histolyticum* also causes serious tissue infections like gangrene.

You may be wondering why spores of *Clostridium tetani* or *Clostridium botulinum* weren't used instead of spores of *Bacillus anthracis*. There are two reasons. First, most people have been immunized against tetanus, so that organism can be ruled out. Second, members of the genus *Clostridium* can't grow in environments that have oxygen, so they wouldn't have been able to grow in the victims' lungs. These spores would have had the best chance of growing if they'd been stabbed into the victims' bodies, like *Clostridium tetani* spores getting into your foot when you step on a nail.

Viruses and Virologists:

Let's move on to **viruses**. Viruses are often referred to as **particles**. You'll never hear a microbiologist or **virologist** (a microbiologist who specializes in viruses or specializes in virology), talk about virus cells. Viruses are not cells. Viruses can't do the normal cell activities of eating and drinking, making cell proteins that act as enzymes and construction materials, or reproducing themselves without help from a cell. Because viruses need a cell to play host to them and help them reproduce themselves, they are termed **obligate intracellular parasites**.

Viruses can survive for a short while outside a cell, just as a fish can survive for a short time out of water. How long the viruses survive outside the host cells varies with the kind of virus and the environmental conditions. Some viruses can only survive a short time. Others can survive for hours. UV light and temperature extremes are environmental conditions that reduce the time a virus can survive.

Human, plant, animal and bacterial cells have **both** DNA and RNA. Viruses have only DNA **or** RNA, but **never** both. The virus DNA or RNA is called the virus **genome**. If the organism has both DNA and RNA, by definition, it isn't a virus.

The virus DNA or RNA is housed inside a **protein shell** called a **capsid**. The protein shell may be a patchwork of smaller pieces of protein. Those smaller pieces are called **capsomeres**. Some viruses have an **envelope**, a wrapping of its host cell's membrane outside the capsid. The viruses that have envelopes take a piece of the host cell's membrane with them when they exit the host cell.

Viruses don't have the ability to reproduce themselves the way human, plant, animal and bacterial cells do. The nonscientific description of how a virus reproduces itself is that it **hijacks** the host cell.

Imagine the host cell is a car and the virus is a "car-jacker." A human car-jacker uses a gun or other weapon to force his way into the car. A virus uses an enzyme in place of the gun. Once inside the car, the human car-jacker uses the gun to force the driver of the car to go where he wants. The virus uses another enzyme to insert its nucleic acid onto the host cell's DNA, to control host cell and force the cell to make more viruses. The hijacked host cell makes virus genomes and virus capsids. The genome is inserted into the capsid while it is inside the host cell, then the completed virus exits the host cell. Many times, the host cell is broken open to release the viruses—sort of like the car-jacker causing a wreck during a high-speed chase and running away from it. Enveloped viruses take a piece of the host cell membrane as a souvenir, like the car-jacker stealing the driver's wallet.

After a virus hijacks the cell and forces the host to make virus particles, the virus particles break the cell. This releases the virus particles made in the cell. Those particles will go on to infect more cells. The cycle repeats until the host's immune system fights off the viruses and stops them from hijacking more cells or the host organism dies.

What determines the kind of cell that is the **host** to the virus? The host cells are those cells that have the enzymes the virus needs to reproduce itself. Some viruses have only a few hosts, such as the virus that causes canine distemper. Others, such as a rabies virus are less picky. Any warm-blooded animal, including humans, can be a host to rabies.

You know there are **human viruses**. You've been infected by some of them. Every cold is caused by a virus. Viruses infect other animals, as well. There are also **plant viruses**, such as the tobacco mosaic virus. Tobacco mosaic virus was the first virus studied in the lab. Plant viruses are the cause of streaks on leaves and petals. They weaken the plants.

Other viruses infect **insects**. Some of these viruses are so host specific that they can be used as insecticides to kill off locusts, without hurting the good insects, such as the ladybug.

Even **bacteria** can be infected by viruses. Viruses that infect bacteria are often called **bacteriophage** or simply **phage**. Phage are important, because many bacteria don't cause human disease unless they are infected with phage. Diphtheria is caused by the bacterium, *Corynebacterium diphtheriae* that is infected with phage. Without the phage, the bacterium is harmless. Botulism toxin is only produced by *Clostridium botulinum* bacteria that are infected with phage.

Phage are referred to as **lytic** and **lysogenic**. Lytic phage hijack the bacterial cell like the car-jacking analogy given above.

The non-scientific example for how lysogenic phage operate is: Imagine that a taxi represents the bacterial cell and the taxi driver represents the bacterial genome. The passenger gets into the taxi and tells the driver where to go. The passenger is like the lysogenic virus. The passenger asks the driver to take him to the airport. The driver complies. Driving to the airport at the request of the passenger is like a *Clostridium botulinum* bacterium producing botulism toxin at the "request" of the lysogenic virus. Lysogenic viruses don't damage the host cells, just as a passenger in a taxi doesn't wreck the taxi the way a car-jacker may wreck the hijacked car. Lysogenic viruses are reproduced when the host bacterium reproduces and they remain inside the host cell.

Prions are a subtype of the virus category. Prions consist of **protein only**. Prion diseases have been in the news. Mad cow

15

is a prion disease. There are prions that attack other animals, including humans. All prion diseases are what is known as "spongiform encephalitis." Spongiform encephalitis is characterized by degeneration of the brain cells so that the brain resembles a kitchen sponge.

There are now a few antiviral antibiotics. Antibiotics such as penicillin, that treat bacterial infections, are useless against viral infections. Antibacterial antibiotics have been given to patients who have had virus infections, which has contributed to bacteria becoming resistant to antibiotics.

Fungi and Mycologists:

The next type of germs are **fungi**, singular **fungus**. Fungi are further divided into the categories of **filamentous fungi**, best known as **"mold"** and the second category, **yeasts**. You probably know there are other fungi, such as mushrooms, but they aren't categorized as germs. Fungal cells have many of the same characteristics as plant cells, including the sub-cellular organelles, a nucleus and a cell wall.

Filamentous fungi cause athlete's foot, toenail infections and lung infections, ring worm. The yeast form of fungi causes vaginal yeast infections and a mouth infection called thrush.

The fungi have many of the same cell structures as human and animal cells. Because the fungi have many of the same cell structures as human and animal cells, including a nuclear membrane, antibiotics that treat bacterial infections are useless against fungi.

The study of fungi is called **mycology**. The people who study fungi are **mycologists** or, of course, microbiologists if you want to keep the general term.

Parasites and Parasitologists:

The next type of germ is the group called parasites. Parasites are animals, so their cells have the same structures as human cells and other animal cells.

Parasites are divided into the subgroups **protozoans** (1 celled animals) and **helminths** (worms). If you have ever seen a tape worm in a jar of formaldehyde in a biology lab, you know there's nothing microscopic about the worm. Tape worms can be many feet long. The eggs of the worms are microscopic, as are the protozoans, hence the inclusion of parasites into the general category of germs or microorganisms.

Scientists who study parasites are **parasitologists**. The study of parasites is called **parasitology**. Those parasitologists who specialize in the worms are called **helminthologists**, because they study **helminthology**. Those who specialize in the protozoans are **protozoologists** and study **protozoology**.

Some of the diseases caused by parasitic protozoans are giardiasis and amoebic dysentary. Helminth infections include trichinosis, heartworm that infects pets, tape worm, pin worm and round worm infections.

As you might expect, antibiotics that are active against the parasites are very different from the antibiotics used to treat bacterial, viral or fungal infections. These antibiotics are more likely to be mildly toxic to the patient, because both patient and germ are animals.

Algae and Phycologists:

Finally, **algae** are included in the general classification of germs. There is only one known species of algae that causes disease and it is not a common disease. Algae are better known for growing in your fish tanks and as food for marine life.

The study of algae is called **phycology**. As you can probably guess, the person whose scientific specialty is phycology is called a **phycologist**.

Other specialists:

Epidemiologists are the people who study the statistics of disease. They are the people who keep score on the flu epidemics every year. Epidemiologists are also medical detectives, because they search for the source of the disease, such as the Anthrax spores in the letters sent in the Fall of 2001.

Molecular Biologists study genomes. They may work with viruses or the genetic engineering of antibiotics.

Good Germs:

As indicated earlier, not all microorganisms cause disease. Many microbes give us food, such as yogurt, sour cream, cheese of all kinds, beer, wine, sauerkraut, and yeast bread. All the aforementioned foods are produced by yeasts, bacteria and molds. The microbes may also be eaten as they are, such as the brewer's or nutritional yeast that some people eat to get extra B-complex vitamins and protein.

The **"active cultures"** in yogurt are live bacteria that ferment the lactose in the milk into lactic acid. The lactic acid curdles the milk into the thick yogurt. There is evidence that the active cultures in the yogurt help prevent disease. You have probably seen some advertising on television for yogurt-type foods that contain cultures that the ads claim will keep you healthy. This isn't a new concept. In the 19th century, a microbiologist named Eli Metchnikoff suggested that eating food that contained certain bacteria, kept people healthy and allowed them to live longer than people who didn't get those bacteria in their food.

Other microbes produce **antibiotics**. Penicillin is produced by the mold *Penicillium notatum*. "Mycin" antibiotics, such as erythromycin, are produced by the *Streptomyces species* of bacteria. Other molds and bacteria produce antibiotics as well.

Oil spills are cleaned up by bacteria that eat the oil. Many other pollutants and insecticides are broken down to harmless substances by bacteria.

Bacteria digest the hay that cows eat and excrete substances that nourish the cows. Bacteria in your intestines produce some vitamins that you need. Those same bacteria protect you from disease producing germs, by keeping the disease-producers from taking over your body.

Spirulina sold as a food supplement in health food stores is a kind of algae. Other kinds of algae may be eaten by humans as well as by marine life.

Where do germs live?

The answer to that question varies by the category of germ. Bacteria live almost everywhere. Some bacteria live in the mineral hot springs at Yellowstone National Park or around thermal vents at the bottom of the ocean. Other bacteria live in extremely cold places. Some bacteria like high salt concentrations. Still others like places that don't have any oxygen. Some bacteria that like places that have no oxygen, live in one of the stomachs of cattle and sheep, the stomach called the rumen.

Bacteria can be found on any surface in the world, which makes it hard to create sterile (no germs), conditions in an operating room. When I say any surface, that also means you. Bacteria are on your skin and hair, in your mouth, on your clothes. Bacteria are on anything you might touch during the course of the day and many of those bacteria can cause disease if they get the chance. This is why you were always told to wash your hands before you eat.

Fungi can be found most places, though they are more picky than bacteria. Fungi require oxygen, so you won't find them growing in the places that have no oxygen. They also like more moderate temperatures, so they don't grow in around the thermal vents in the bottom of the ocean. Fungi, molds in particular will grow well in your refrigerator, though. Fungi may also be on your body.

Viruses can only reproduce themselves inside a living cell, but they can survive outside a cell for varying lengths of time. They may be on any surface, such as a doorknob just waiting to give you a cold or the flu. Sunlight kills viruses fairly quickly, which is part of the reason why flu epidemics don't occur during the summer when people are out enjoying the sun.

Parasites like it warm and wet. Parasitic diseases are more common in warm, humid locales than in cold places. **Algae** like water. Just look at the sides of a fish tank to verify this.

Chapter Three
They get around or How germs spread

Some germs spread through the air. They may spread like dust, as in the case of the anthrax spores. Other germs spread in a mist, like the spray cleaner you use on your windows. This form of spread is called **air droplet**. When you cough, sneeze or just talk, you produce a fine mist composed of the moisture in your mouth and any stray bacteria or viruses that might be in your mouth. Cold and flu virus, smallpox virus and the bacterium that causes strep throat are examples of the germs that spread as droplets.

Other germs are spread by **direct contact** with something that is contaminated with the germ. Examples are hepatitis and AIDS that are spread by direct contact with blood, body fluids or hypodermic needles that are contaminated with the germs. Many germs that cause diarrhea or other intestinal distress are spread by direct contact with feces, or **sewage contaminated food or water**. Eating or drinking contaminated food and water is another form of direct contact. Wound infections can be spread by direct contact with used bandages. **Sexually transmitted diseases** are spread by direct contact with someone who is already infected with the disease. Tetanus can be acquired by direct contact with something that is contaminated with the germs and causes a puncture wound. A

22

special type of direct contact is the **nosocomial infection**, in which germs are transmitted from contaminated medical equipment to the patient.

Insects can spread germs. Malaria, Western Equine Encephalitis, Venezuelan Equine Encephalitis and West Nile Fever are examples of diseases that are spread by mosquitoes. Typhus and plague are spread by fleas. Plague may also be spread by droplets when it takes the pneumonic form (pneumonia). Colorado Tick Fever, Rocky Mountain Spotted Fever and Lyme disease are spread by ticks. Spread of disease by insects is called an **insect vector**.

Insects and animals can also spread germs simply by transporting them from one place to another. An example of this is when a fly lands on dog feces, then lands on food. Germs are transported on the fly's feet. Another example is when a cat scratches in a litter box, then walks around on the kitchen counters.

Some germs can be spread from the mother to the fetus across the placenta. Babies born with AIDS are an example of this kind of spread of disease. This type of spread is referred to as **vertical or trans-placental**. The ways disease spreads listed above are referred to as **horizontal**.

Try this at home or in your classroom

Shaking hands with your next cold: to demonstrate how far a cold virus can spread. You'll need a **bowl of water** and a dozen or more people. The steps are:

1. Have the people line up.
2. The first person will dip his right hand into the bowl of water.
3. Without shaking off the water, person #1 shakes hands with person #2, which wets person #2's hand.
4. Person #2 then shakes hands with person #3, wetting person #3's hand and so on down the line until the water is no longer passed to the next person.

The water represents the cold virus that would be on a person's hand if he used the palm of his hand to block a sneeze. Everyone whose hand got wet would represent someone who was infected with the cold virus from a person's hand. Depending on how fast the water evaporated, six or more people's hands got wet in this demonstration.

First variation—Quarantining the "infected person": Now, try the handshake experiment with a twist.

1. Person #1 will dip his hand in the bowl of water.
2. Person #1 will walk over to stand in the "designated quarantine corner" without shaking hands with anyone.

3. The other people shake hands.

Note whose hands got wet from the water person #1 dipped his hand into. No one else's hand got wet. This is what happens when you get a cold and stay home from work. You're not sharing it with anyone at work, though you might be sharing it with the people at home.

Second variation—Quarantining the "healthy people": Repeat the handshake experiment as follows:
1. Person #1 dips his hand into the bowl of water.
2. Without of shaking hands with person #1, all the other people go over to stand in the "designated quarantine corner."

Whose hands got wet this time? This version represents the reverse method of containing an infectious disease, also known as a communicable disease. This method protects the uninfected people by keeping them away from the germs.

This method is used in hospitals to protect burn patients, transplant patients and trauma patients from the germs they might get from other patients. Canceling classes when there is a flu epidemic stampeding though a school, is another way this method can be used to protect uninfected people from getting sick. If you're a fan of Tom Clancy's novels, you'll probably remember that this method showed up in *Executive Order*, during the Ebola epidemic.

Tracing the Infection in real life:

The shaking hands demonstration is a simple example of how a cold virus can spread. In real life, handshakes aren't the only way that a cold virus spreads. The real life spread would be more like the following:

1. Student A has a cold and goes to school. He was taught to cover his mouth and nose with his hand when he sneezes. He sneezes and coughs, covering his mouth with his right hand, contaminating his hand.

2. Student A uses his right hand to grab the door handle when he goes into the building that houses the bursar's office. This leaves cold viruses on the door handle. For the next several hours, anyone who touches the door handle after him will pick up the cold virus.

3. Student A goes to the bursar's office to pay his tuition bill. He takes his credit card from his wallet with his right hand and gives the card to the bursar. In the process, he hands the bursar some cold viruses.

4. The bursar runs the credit card through the machine, then hands it back to Student A along with the receipt to be signed. Student A borrows a pen to sign the

receipt, contaminating the pen. His hand also contaminates the counter in front of him. Anyone who touches the counter will become contaminated with the cold virus for several hours after Student A was there.

5. The bursar rubs his eye with his contaminated hand, putting the cold virus into his eyes. The virus will travel down the tear ducts to the back of his throat and nose and invade some cells, starting a cold.

6. Student B comes to the bursar's office from the registrar's office. He uses the contaminated pen to write his tuition check. His hand becomes contaminated with the cold virus. He picks his nose and infects himself with the cold virus.

7. Student B is student teaching. He takes the cold virus into the grade school where he is assigned. He leaves cold germs on the door handles, water fountain handles and chalk holder while he is there.

8. If a member of the custodial staff touches a tissue that was used by someone who has a cold, that person will become infected from the germs on the tissue. And the virus continues to spread...

One person with a cold can potentially infect hundreds of other people with his cold if he goes to work or to school. If Student A had stayed home with his cold, he wouldn't have infected anyone at the college or the grade school where Student B is student teaching.

Stopping the spread of a disease:

The handshake exercises demonstrated how quarantine works. If you're a student of history, you may recall reading about homes being **quarantined** when the people living there had smallpox, plague or other highly contagious diseases.

If you stay home when you have a cold, you are voluntarily quarantining yourself. You can help stop the spread of your cold, even when you have to go out. Other methods for stopping the spread of a cold include:

1. Cover your mouth and nose to block your coughs and sneezes.
2. Have a tissue in your hand, instead of sneezing or coughing into your bare hand.
3. Use your upper arm to block a sneeze or cough instead of your hand. Your arm will be contaminated with cold virus, but you don't touch as many things with your arm as you do with your hand.
4. When you have a cold, wash your hands frequently with plain, old soap and water.

To avoid becoming infected with a cold or the flu wash your hands often. If you're around someone who has a cold, try to avoid rubbing your eyes until you've washed your hands. Soap and water doesn't kill the germs, but it does wash them down the drain, and that is all you need. Rinsing your hands with water alone will remove some of the germs, though soap will remove germs that may be caught in skin oils on your hands. You don't need antibacterial soap, because colds aren't caused by bacteria.

Hand washing 101:

How important is hand washing for avoiding infection? Consider the following examples:

Scenario #1: Imagine that you are a prison guard standing alone among 100 unruly inmates. You're armed. All the inmates decide to attack you at the same time. Even if your gun is fully automatic and will shoot continuously as long as your finger is on the trigger, the odds are that at least one of the inmates will get you.

Scenario #2: Imagine that you are a prison guard standing with 1 inmate. If this inmate becomes unruly, you have a better chance of subduing the prisoner, even if you are not armed.

Washing your hands is like removing 99 of the prisoners. If you only have 1 germ on your hands, you have much better odds that you won't get sick than you have if you have 100 germs on your hands.

Correct method for washing your hands:

1. turn on the water, soap and rinse your hands as usual

2. turn the water off *without* touching the taps. Some faucets are turned on with a lever. You can press the lever down with your wrist. If you have to twist the tap to turn it on, you can soap the handle as well as your hands, or use a paper towel to turn the water off.

3. dry your hands. If you are using a paper towel, you can use your wrist to roll the towels down, then grab the towel and dry your hands.

Public Restrooms 101

Turning on a faucet with your germy hands leaves germs on the faucet. People who don't wash their hands leave germs on the door handle. So, how do you protect your clean hands when you turn off the water faucet in a public restroom and open the door to exit?

Many public restrooms have water faucets that turn on automatically when you wave your hand under the tap. If you have to turn the water on and off yourself, use your wrist or elbow if you can. If you have to use your hand to turn the water off, leave the water running while you grab a paper towel, dry your hands, then use the towel to protect your hand when you turn off the water.

Keep the towel to protect your hand when you open the door and leave the restroom. Discard the towel in a trash can outside the restroom. If the restroom door pushes open, you can use your elbow, shoulder or tush to push the door open.

Does this sound like overkill? Possibly, but you'd be completely grossed out if you knew what germs are routinely found on door handles, stair rails, elevator buttons and anything else that people frequently touch in public places.

Avoid doing this at home and everywhere else.

Whipping up a little chicken Caesar and Salmonella salad for the family reunion:

Imagine that you're taking chicken Caesar salad to the family reunion picnic. You start by chopping chicken breasts into strips for cooking. In the process, some chicken blood gets on your cutting board and on the counter under the cutting board. This chicken happens to be contaminated with *Salmonella*.

Cutting the raw chicken contaminates your hands, the knife, the cutting board and anything you touch, with *Salmonella*. You know that your hands, the knife and the cutting board are contaminated with any germ that is contaminating the chicken. You intend to wash your hands, the knife and board as soon as you get the chicken into the skillet and started frying.

Do you need to be concerned about *Salmonella* being on anything else in your kitchen? Yes. You've touched the knob that controls the burner on your stove, the faucet handle and probably the handle of the skillet and the spatula you used to spread the chicken in the skillet. The counter where you worked was also contaminated by blood dripping off the cutting board.

If you're like most people, you'll touch the water tap to turn on the water to wash your hands, then you'll touch the tap again to turn the water off. This re-contaminates your hand with *Salmonella*. When you tear the lettuce and chop any other vegetables for the salad, you'll be transferring *Salmonella* from your re-contaminated hands to the vegetables.

Now, you put the chopped and contaminated vegetables into the salad bowl and top them with the cooked chicken. Again, if you're like most people, you'll leave the salad out while you make the Caesar dressing.

You get a couple of eggs, Parmesan cheese and a few of your secret ingredients out and set them on the counter. You break the eggs into your mixing bowl. A bit of shell gets into the bowl. The shell is contaminated by your hands and carries the *Salmonella* into the egg mixture. The egg shells may be contaminated with *Salmonella*, even if your hands aren't.

You pour the dressing on the salad, cover the bowl, and leave it on the counter, ready to carry out the door as soon as your spouse and kids are rounded up. Ten minutes later, you and your family take the salad out to the car and head to the picnic. You arrive at the park 20 minutes later. You put the salad bowl on the buffet table with the rest of the food. People fill their plates and eat.

Some people go back for seconds on the salad. Twenty-four to thirty-six hours later, all the people who ate the salad are sick with food poisoning. Unfortunately, everyone who ate your cousin's *Salmonella* and banana cream pie gets sick, too.

Could the food poisoning have been prevented? Yes. The following changes in the procedure could have prevented it:

1. Avoid contaminating the raw food by washing the counter top where you cut the chicken before cutting anything else, even if it doesn't look like it got any chicken blood on it.

2. Don't touch anything until you've washed your hands. You can avoid touching the water tap with your contaminated hands by any of several methods.

   a. If your water tap can be turned on with your wrist or a clean spoon, leave a spoon handy and turn the water on that way.

   b. fill your sink or a dish pan with hot, soapy water before you start cutting the chicken. Use that water to wash your hands when you've finished cutting the chicken so that you don't have to touch the water taps.

   c. have someone else turn the water and stove on for you.

   d. have a pump bottle containing waterless hand sanitizer or plain rubbing alcohol in the kitchen for sanitizing your hands. Use your wrist to push the pump down.

e. if you have to turn on the taps with your hands coated in *Salmonella*, wash the tap with soapy water before you turn it off.

3. Keep the salad cold by putting it in a picnic cooler. Leave the salad in the cooler at the picnic or put some plastic bags full of ice around the bowl on the table.

Poop-patrol: Try this at home or while walking your dog.

While you're walking your dog, you're expected to pick up any souvenirs your dog leaves in someone's yard. You can carry a shovel with you or use plastic bags. Here are two ways to use the bags:

Method 1: You're using a zipper-sealed bag, so you can seal the bag after you've picked up what your dog leaves behind.

When using a zipper-sealed bag to pick up dog feces, turn the bag inside out and slip it over your hand like a mitten. Pick up the poop, then pull the top of the bag over your hand and the poop, so that you have the souvenir in the bag and your hand outside of the bag. Zip the bag closed and dispose of it in the normal manner.

Method 2: You have two bags, one to use as a container and the other to use as a protective "glove." With this method, you'll slip the "glove" bag over your hand, pick up the poop and put everything into the carrier bag. When you pull your hand out of the carrier bag, leave the "glove" bag inside.

Practice everyday prevention

I'm sure you've noticed that the examples on how germs spread are ordinary, rather than focusing on what to do in the case of bio-terrorism. There's a reason for this. We are all vulnerable to colds and food poisoning all the time. We're all expected to pick up what our dogs leave in someone's yard. If you know what to do when confronted with the ordinary germs, you'll know what to do if there is a bio-terrorism alert.

The "shaking hands with your next cold" exercise shows how someone can spread a cold or the flu. It is also how someone could spread smallpox or pneumonic plague or other airborne disease. For the majority of us, the important difference between a cold and smallpox is the lethality of the disease, not what we need to do.

The methods you can use to avoid smearing *Salmonella* around your kitchen, work just as well to avoid leaving a trail of more serious disease germs around your house. These methods are also good for preventing spread of hepatitis A, which is commonly spread by food handlers.

If you practice safe handling of contaminated materials by doing your "poop-patrol" duties, you'll be able to pick up a suspicious package with less risk of contaminating yourself.

Using the bags over your hand keeps the germs off your hands and it keeps your fingerprints off the package. Changing your shirt and washing your hands after the package is bagged, reduces your risk of being contaminated even more.

You may choose not to handle a suspicious package if you ever see one. If you can, you may want to lay a sheet of plastic wrap over the package until the proper authorities can get there and take care of it. The sheet of plastic wrap will help keep anything that could leak out of the package from becoming airborne.

If there ever is a bio-terrorism alert, the local health department or **Centers for Disease Control (CDC)** will broadcast instructions for protecting yourself and your family. You'll want to follow those instructions as well as practicing safe handling of infected materials and the other safety procedures for preventing ordinary illnesses.

Chapter Four
Toxins and enzymes

Not only are germs found everywhere, they have weapons they can use against you. The first weapon germs have to use against you is the ability to **attach to your cells**. Attaching to your cells is the first step in infection. This puts the germs in the proper positions to utilize their other weapons. The second assault germs will launch against you, is to release **enzymes** or **toxins**.

Enzymes

Germs may release enzymes that digest pieces of your cells or tissues. Different germs have different enzymes.

One of the enzymes that germs may produce is **lecithinase**. Lecithinase digests the membranes of your cells.

Another enzyme that germs may produce is **deoxyribonuclease**. This enzyme digests your DNA.

Collagenase is another enzyme that some germs may produce. It digests collagen. Collagen is a protein that is in a fiber form. Collagen is a structural protein found in hair, skin and other parts of your body.

Some germs produce an enzyme called **coagulase**. Coagulase causes your blood to clot. When your blood clots around the germ that is producing the coagulase, your white blood cells and antibodies can't get to the germ to destroy it. You'll learn more about antibodies and white blood cells in the chapter "Defending Yourself Against Germs."

Other germs produce an enzyme called **streptokinase**. Streptokinase dissolves blood clots. Streptokinase allows germs to get into your blood stream and circulate through your body.

NOTE on Streptokinase: You may recall hearing news reports about a drug called streptokinase that is used as a **"clot-busting" drug** to treat heart attacks and strokes. The drug called streptokinase is the enzyme that some bacteria produce.

Some germs produce an enzyme called **hyaluronidase**. Hyaluronidase digests hyaluronic acid. Remember that hyaluronic acid is a substance that cements your cells together to form tissues. Hyaluronidase is one of the enzymes that allows the germs commonly known as "flesh-eating" bacteria, to eat their way through your body. The scientific name for flesh-eating bacteria is *Streptococcus pyogenes* group A (this grouping is based on the antibodies that fight this germ).

Toxins

Toxins are another weapon that germs may use against you. There are two general categories of toxins. These categories are: **endotoxins** and **exotoxins**.

Endotoxins

Endotoxins are part of the cell walls of some bacteria. Usually, you don't develop antibodies against endotoxins. Endotoxins are not destroyed by boiling. Another name for endotoxin is **pyrogen**.

Pyrogens get their name from the fact that they cause a **fever** if they get into your blood. Pyrogens are a major concern for medical equipment and pharmaceutical manufacturers and, believe it or not, electronic equipment manufacturers.

As you can probably guess, the fact that pyrogens cause fevers is why medical equipment and drug manufacturers are concerned about pyrogens. If pyrogens are on or in intravenous tubing or fluids and get into the blood of a patient who might already have a fever, they'll cause the fever to go higher. This could send the fever to a fatal level.

Pyrogens are destructive to electronic equipment if they are in the water used to wash micro-circuitry. The pyrogens cause short-circuits on the microcircuit.

Some pyrogens are found in tap water. These don't cause you problems if you drink them. Tap water must be specially treated to remove the pyrogens before it is used for washing medical equipment or electronic parts or for making liquid medications and intravenous fluids. Filters are often used to remove pyrogens from tap water.

Killing the bacteria doesn't remove the pyrogens. Dead bacteria in intravenous fluids are just as pyrogenic as live bacteria. Dead bacteria cause short-circuits on microchips, too.

One of the tests for pyrogens uses a small amount of an extract from the blood of the horseshoe crab, *Limulus polyphemous*. When the blood extract from the horseshoe crab mixes with pyrogens, it clots. Ironically, this was discovered as a result of sewage pollution of coastal water.

Horseshoe crabs donate a small amount of blood and are returned to the ocean. They aren't harmed any more than a human is by the blood donation.

Exotoxins

Exotoxins, the other type of toxin, are proteins. They are excreted by the germs that produce them. Exotoxins are destroyed by heat, such as boiling for 20 minutes. Because exotoxins are proteins, they are good antigens. Good antigens, means that you develop antibodies against them.

Two of the best-known exotoxins are **botulism** and **tetanus** toxins. Other germs produce exotoxins, too. Toxins cause some of the symptoms of anthrax and diphtheria. You'll learn more about the germs that produce exotoxins in the chapter on "Bacterial Diseases."

Notes on Botulism toxin:

Botulism toxin is produced by the bacterium *Clostridium botulinum*. It is one of the deadliest toxins known. Because botulism toxin is an exotoxin, heat destroys it. If you are exposed to botulism toxin, you can be treated with **antitoxin** (antibodies against the toxin), for botulism.

It is possible to be immunized against botulism toxin. Cattle are immunized against botulism toxin in some geographic areas. There are some experimental toxoids being used to immunize laboratory workers against botulism.

Is there a plan in the U.S. to have mass inoculations for botulism? Not according to the news. The news reports on botulism as a terrorist weapon say that antitoxin is being manufactured.

Is botulism as a weapon of bio-terrorism the greatest risk from botulism that you face? Probably not. A major risk from botulism is from home-canned food, particularly vegetables that were not pressure cooked during the canning process. This is why your grandmother always told you to boil your canned vegetables for 20 minutes before you eat them. Boiling for 20 minutes will destroy the toxin.

Commercially-canned food can also be infected with botulism, but it is *highly unlikely* due to the processing that is done by the commercial food-canners. Their processing methods are so strict, that the odds of a can of food containing botulism is 1 in 1,000,000,000,000 (that's a trillion).

Boiling commercially-canned food for 20 minutes never hurts, but it's a fairly safe assumption that it isn't necessary. If you are going to serve something like "3-Bean-Salad" *always* make it with commercially-canned beans or cook the beans for the salad.

You can also get botulism the same way you can get its relative, tetanus. This is from a wound being infected with the organism. As with tetanus, you're more likely to develop wound botulism from a puncture wound, because oxygen doesn't get into the infected tissues.

On occasion, infants develop botulism. The theory on how they get botulism is that they have the organism in their intestines and it produces toxin that affects them.

Botulism toxin can also be inhaled. If it is used as a biological weapon, this is one way you may be exposed to it. It could also be put into the water supply, but if that happens, you can boil your water for 20 minutes to destroy the toxin.

You've heard news reports about botulism toxin being manufactured in Iraq as a weapon. You may find it interesting to know it's being manufactured in the U.S. for medical and cosmetic use. **BOTOX**, the trademark name of botulism A toxin manufactured by Allergan, is used to treat some medical conditions, such as dystonia. The toxin attacks the nerves that cause muscle contraction and prevents the contractions that characterize these conditions. BOTOX is also used by some people to relax their facial muscles to reduce wrinkles.

Some people who have had the BOTOX treatment for wrinkles report that they have had fewer migraine headaches since they've had the BOTOX treatment.

Chapter Five
Defending Yourself Against Germs

Germs are everywhere. They produce enzymes that digest parts of your cells. They produce toxins that can kill you. Are you helpless against germs? No. If we were helpless against germs, the germs would have killed all of us a long time ago.

Non-Specific Defenses

Your skin, the mucous membranes lining your nose, throat, stomach, intestines and every other body opening, provide a **barrier** to germs. Body fluids, such as tears, digestive juices, urine, mucous and saliva **wash germs away**. In addition, many body fluids contain an enzyme called **lysozyme** that attacks the cell walls of bacteria. Body fluids may contain antibodies that attack invading germs. Stomach acid destroys many germs.

Fever is a defense against germs. Many germs grow best at normal body temperature for humans. Fever raises your body temperature, so the germs don't grow well. High fevers are dangerous to you. They can cause convulsions. Extremely high fevers can be fatal.

Taking aspirin, ibuprofen or acetaminophen to reduce your fever will reduce your risk of convulsions and make you more comfortable. Unfortunately, reducing your fever makes the germs more comfortable, too.

If your child has a fever from a cold, flu or chicken pox, *never* use aspirin to reduce the fever without checking with your doctor first. There have been rare occurrences of Reye's syndrome resulting from a child being given aspirin with a cold, flu or chicken pox.

Despite your external defenses, germs may get into your body. They may gain access through **breaks in your skin or mucous membranes**. They may also gain access by using their toxins and enzymes. If germs get into your body, there are chemicals and special cells that exist specifically to protect you from the invading germs.

The protective cells are the various types of **leukocytes (white blood cells)**. They may surround the invading germs at the site of entry and keep them from going further into your body's tissues. Other leukocytes eat the invading germs. These leukocytes are called **phagocytic cells**.

Some invading germs are able to fight back against the leukocytes. These germs may produce chemicals that poison the leukocytes or digest the leukocytes.

Specific Defenses

Antibodies, also called *gamma*-globulins or immune-globulins are important protective chemicals produced by your body. Antibodies are small chains of protein. These small chains of protein are called **polypeptides**.

Links that make up a protein chain are called **amino acids**. If you use a protein supplement, its label will show a list of the amino acids supplied by the protein supplement.

Your body can make some of the amino acids it needs for making proteins. Other amino acids have to come from the food you eat. These amino acids are called **essential amino acids**, because it is *essential* that you get them from your diet.

The best sources of essential amino acids are milk, meat, fish, or poultry. People who eat vegetarian diets need to be careful to get enough essential amino acids in their diets, from a mix of legumes, nuts and grains.

One of the reasons you need to make sure your diet supplies the essential amino acids, is that the essential amino acids are needed to make antibodies. If you don't get enough essential amino acids in your diet, you won't be able to make antibodies. The essential amino acids are needed for other proteins, as well.

The immunization process

The first step in developing **immunity** to a germ is for the germ or its toxin to get into your body. This can happen naturally when you get sick or artificially when you're given a **vaccine**. Flu shots and tetanus shots are examples of vaccines you might be given.

Some vaccines are dead germs. The **Salk polio vaccine** is an example. Other vaccines, such as the **Sabin polio vaccine** are attenuated live germs. Still other vaccines are **toxins** that have been treated with formaldehyde to destroy the toxicity of the toxin. They are called **toxoids**. Although the toxicity of the toxoids has been destroyed, they still have their antigenic properties. The tetanus immunizations you've received are tetanus toxoid.

Once the germ or toxoid is in your body, your body checks to see if it is part of you or a foreign invader. When your body decides the germ or toxoid is foreign, your immune system begins to make antibodies against the foreign invader.

Foreign invaders that cause your body to create antibodies are called **antigens.** Anything that causes your immune system to create antibodies is said to be antigenic.

Proteins are good antigens. Fats and sugars are mediocre antigens.

Antibodies are **specific** for an antigen. One way to look at this is to imagine that every red car in the world can be unlocked by the same key. Ditto for blue cars being opened by the same key. The key that unlocks all the blue cars, won't unlock any red cars and vice versa.

Let's say that the red cars represent this year's strain of flu virus. The keys that unlock the red cars represent the antibodies against that strain of flu virus. The keys may have been made in response to a flu shot or in response to getting the flu.

Now, let's say that the blue cars represent tetanus toxin or toxoid. The keys that open the blue cars represent the antibodies you developed against the toxin or toxoid. The difference between red and blue cars is obvious. They represent two different germs or toxins, so you wouldn't expect their antibodies to be interchangeable.

If red cars represent flu virus, you may be wondering why you need a flu shot every year. The analogy for this, is to imagine that all red cars manufactured before the year 2000 have one key and all red cars manufactured after the year 2000 have a similar, but different key. The locks have changed in the newer cars. The changed locks are a mutation in the post-2000 red cars. This is what happens with flu viruses. They are still similar, like the red cars, but they are different enough to require different flu vaccine, like the cars with changed locks that require different keys.

Would it be possible to immunize people against colds? Yes, it's possible, but not practical. There are dozens of different viruses that cause colds. Those viruses are mutating all the time, making still more different cold virus antigens. Trying to be immunized against colds is sort of like trying to get the key to every car in the world, when every car has a different key.

Protecting yourself against colds requires more mundane methods. Some of those methods are: eating healthy food, exercising, getting enough rest and....washing your hands frequently with plain soap and water.

Chapter Six
Germ-killing 101

As suggested before, your first line of defense against spreading many diseases is washing your hands. This removes the germs and washes them down the drain. Plain soap is adequate for this job. You don't need to kill the germs before you wash them down the drain.

Sometimes you do need to kill the germs. The general term for killing *all* the germs on or in an item is **sterilization**.

There are many ways to kill germs. Some methods can be used by anyone, anywhere. Other methods require a doctor's prescription. The methods that require a prescription are covered in the chapter on antibiotics. Still other methods require specialized equipment or chemicals and are used in hospitals, laboratories and medical and pharmaceutical manufacturing plants.

Some of these methods kill most of the germs. Other methods kill *all* the germs, or sterilize.

Use anywhere methods:

Heat

Heat is a method of killing germs that we've all heard about and probably used. **Boiling** is a common method of killing germs. You've used this method when you've boiled the jars you plan to use for canning foods, or for sterilizing baby bottles (at least you've done this if you're as old as I am).

There's a variation on boiling that is called **Tyndallization**. Tyndallization is boiling something, allowing it to sit (incubate) for 24 hours, then boiling it again and allowing it to sit again. The reason why this method may be used, is that boiling doesn't kill bacterial spores. The first boiling kills off the vegetative bacteria, then allowing the liquid to sit for 24 hours gives any spores a chance to germinate into vegetative bacteria. The second boiling kills the bacteria that germinated from the spores. Another cycle of incubation and boiling is used in case some spores germinate after the second boiling.

Saturated steam is another heat method you can use. **Pressure cookers** use saturated steam to kill vegetative bacteria *and* spores. This is why you need to use a pressure cooker when you can vegetables at home.

You can also use **dry heat** to kill germs. Baking and frying are forms of dry heat. You may recall hearing that hamburgers need to be cooked until the inside of the burger reaches a temperature of 160° Fahrenheit to kill *E. coli*. If you don't have a meat thermometer, cook the hamburgers until they are no longer pink in the middle, to be sure the meat reached the temperature needed to kill the germs. This doesn't make the hamburgers sterile, but it reduces the number of bacteria.

Incineration is another use of heat for germ killing. Obviously, incineration often destroys the item as well as the germs. You might use incineration for killing the germs in used bandages. If you've ever run a sewing needle through a candle flame before using the needle to pick a sliver out of your hand, you've used incineration to kill germs.

Heat isn't always the best method for killing germs. It isn't practical to boil your kitchen counters or your toilet. Heat may damage whatever it is that you want to sterilize, such as a plastic food cutting mat. Fortunately, there are other ways to kill germs.

Chemicals

Household chemicals used to kill germs are loosely grouped into: **antiseptics** which are chemicals you can use on your skin and **disinfectants**, which are harsher chemicals intended for killing germs on objects, such as medical equipment, counters or toilets.

Isopropyl alcohol in a **70% solution** is one of the best germ killing chemicals. The 70% solution of isopropyl alcohol is the best concentration for germ killing, better even, than 100% isopropyl alcohol. It's cheap and available everywhere. A common name for this germ killing chemical is **rubbing alcohol**. This can be used as an antiseptic or a disinfectant.

Isopropyl alcohol is the germ killing ingredient in the waterless hand sanitizers available on the market. The sanitizers consist of isopropyl alcohol, a thickening agent to make them a gel, a moisturizer and usually some perfume.

Rubbing alcohol kills germs by denaturing proteins in the germ cells. You've seen denatured protein if you've ever fried an egg. The egg white is clear before cooking, then turns to an opaque white color after being cooked. That opaque white color is a sign that the protein in the egg white is denatured.

Denatured proteins don't function. In a bacterial cell, it means that the enzymes don't digest its food or allow the bacterium to make or repair cell structures. The cell structures that contain protein, cease to function as well.

You can use rubbing alcohol on skin, on surfaces, such as tabletop or toilet seats—essentially anything that isn't damaged by moisture, *except food.* If you use rubbing alcohol on cooking utensils, let it evaporate from the utensils, then rinse the utensils several times in running water.

NEVER use rubbing alcohol on food or on sores in your mouth. Rubbing alcohol is as toxic to you as it is to the germs.

Ethyl alcohol, the kind you can drink, is also good for killing germs. If you want to use a type of alcohol to kill germs on food or in your mouth, use your favorite liquor or mouthwash.

Phenol is another common germ killer. It is often found in salves, anesthetic throat sprays and cleaning products. Phenol has the added benefit of being an anesthetic in throat sprays. The older name for phenol is **carbolic acid**, hence the name carbolated salves.

Another readily available germ killer for home use, is **sodium hypochlorite**, commonly known as **chlorine bleach**. Liquid chlorine bleach is a solution of sodium hypochlorite, usually 5.25% or 6%. You can use diluted bleach to kill germs on counters and bathroom fixtures. When you dilute the bleach, **mix 1 cup of bleach into 9 cups of water**. This dilution of bleach is used on counters and other surfaces in labs that work with anthrax.

If you've ever spilled liquid chlorine bleach on fabrics, you know it will bleach the color out of the fabric. Be careful with the bleach to avoid ruining your clothes.

Liquid chlorine bleach can also be used to kill germs in drinking water. If you need to kill germs in your drinking water, the EPA suggests that you use:

10 DROPS of liquid chlorine bleach per quart of water.

Wait 30 minutes, then smell the water. You should be able to smell the chlorine in the water. If you don't smell chlorine in the water, add another 10 drops of bleach to the water and wait 30 minutes again. Repeat this process until you do smell chlorine in the water.

The water will now be safe to drink, but it might not taste very good. You can improve the taste by pouring the water back and forth between two clean containers for a few minutes. Pouring the water back and forth puts oxygen to it.

You can get more information on using liquid chlorine bleach to kill germs in drinking water at the Environmental Protection Agency's web site or the web site of the Chlorox Company, see: **www.epa.gov or www.chlorox.com.**

You may need to use liquid chlorine bleach to treat your drinking water when there is a natural disaster, such as a flood. You can also use the bleach to treat your drinking water when you're out camping.

Hydrogen peroxide is another common household chemical that kills germs. Hydrogen peroxide doesn't kill all germs. Some germs have enzymes that break hydrogen peroxide down to water and oxygen. Hydrogen peroxide is a good antiseptic to use in and on wounds. Hydrogen peroxide can also be used in your mouth, though you might look like a rabid dog—frothing at the mouth when the peroxide foams.

The antiseptics available at the supermarket or drugstore also kill germs. These are intended for use on skin, but not for any sores in your mouth, unless the label specifically says it is safe to use in your mouth. To kill germs in your mouth, mouthwashes containing either hydrogen peroxide or ethyl alcohol or throat sprays with phenol are good choices.

Two household items you may not have thought about as germ killers, are **salt** and **sugar**. Both kill germs by sucking all the water out of the cytoplasm of the germs. This is why beef jerky doesn't spoil and why bacteria don't grow in jams and jellies. Mold can grow on beef jerky or jams and jellies. Salt and sugar are most commonly used in food preservation. Technically you could use salt or sugar to kill the germs in a cut on your hand, but I don't recommend it. If you try it, you'll have a new understanding of the old cliche about rubbing salt into wounds.

Dehydration or **drying** is another method of food preservation that kills germs by sucking the water out of them. Salt and sugar are often used with dehydration to preserve foods, like fruits.

Laboratory and Industrial germ-killing

Heat is used in laboratories, hospitals and manufacturing plants to kill germs. They may use dry heat, boiling, saturated steam or incineration like you could use at home. Labs, hospitals and manufacturing facilities use large pressure cookers called **autoclaves** to sterilize medical equipment and intravenous solutions. The autoclaves may only be large enough to handle a few liter flasks or a few thousand test tubes. They may also be large enough to handle a dozen 4ft x 4ft x 6ft high pallets of product. Commercial canneries use heat to kill off *Clostridium botulinum* spores in the food they are canning.

Pasteurization is another heat method for killing germs. This is usually done by commercial dairies to kill the pathogenic (disease producing bacteria) in milk and other dairy products. Pasteurization doesn't sterilize the milk.

Radiation is one method of killing germs in manufacturing facilities. You've heard news reports about radiation being used to kill the anthrax spores on mail. Radiation can be used on food and medical equipment as well. One drawback to using radiation, is that it can alter the chemical structure of the item you're sterilizing with it. Plastics that are flexible before radiation sterilization may become brittle and fragile after radiation sterilization.

The same **ultraviolet light** that causes sunburns and skin cancer, kills germs. UV light does the same kind of damage to germs that it does to your skin, which is how it kills germs. UV light is often used to sterilize surfaces in rooms such as operating rooms and drug or medical equipment manufacturing areas.

When radiation and UV light aren't practical methods for killing germs, the next choice is a chemical. Laboratories, hospitals and manufacturing plants may use the same **chemicals** to kill germs that are available to you at home, particularly chlorine bleach and alcohol. Labs also use chemicals that aren't available for home use.

Labs often use **aldehydes** to kill germs. You've probably seen frogs or organs preserved in **formaldehyde** in school biology labs. By killing the germs on the organs, formaldehyde keeps the organs from rotting.

Another aldehyde that is used in labs to kill germs is **glutaraldehyde**. This is used for sterilizing objects in labs, such as the counter tops and animal cages.

Ethylene oxide and **propylene oxide** are gasses that are used in some hospital and industrial sterilizers. These gasses kill germs at low temperatures, so they are useful for sterilizing medical equipment that would be damaged by heat or radiation, such as plastic syringes.

A more exotic place you might find ethylene oxide sterilizers, is at NASA. NASA uses ethylene oxide sterilizers to sterilize things that are carried into space, so we don't introduce earth germs onto other planets or the moon. Some state historical societies have ethylene oxide sterilizers for sterilizing items like animal skins that may have picked up anthrax spores from the environment.

There is a common sterilizing agent that doesn't kill germs. This agent is a **filter**. You've seen air purifiers and vacuum cleaners that have **HEPA filters**. HEPA (High Efficiency Particle Arresters) filters are also used to sterilize the air in operating rooms and manufacturing clean rooms. They remove the germs from the air that passes through them. Clean rooms are often as concerned with removing nonliving particles, such as dust, dead skin cells and clothing fibers as they are with removing living germs, because particles damage electrical components that are manufactured in clean room. Bacteria, dead or alive, can cause short-circuits on microchips.

Drug manufacturing often uses filters to remove germs from liquid medicines. Again, this doesn't kill the germs, but the medicines are sterile because there are no germs left in them after they're filtered.

Antibiotics are also chemicals that kill germs. They are covered in a separate chapter.

Home emergency kit

Every home should have an emergency kit that contains:

1. first aid items
2. nonperishable food
3. bottled water
4. flashlights and batteries
5. battery powered radio
6. blankets
7. **rubbing alcohol** and **household bleach**
8. comfort items, like favorite snacks or toys
9. You may want to have a camp stove or barbecue to *use outdoors* to heat your food and boil water if necessary

The primary reason for having such a kit is to be prepared for natural disasters such as floods, storms, power lines being knocked down by wind or snow and any other kind of natural disaster that might disrupt normal electrical services and water processing. A secondary reason for having a home emergency kit is in the event of a terrorist attack that disrupts normal services.

Comprehensive lists of suggested items for an emergency kit are often available from local health departments, television stations, local emergency preparedness offices, and various government agencies. Other sources of lists are companies that manufacture or distribute items that are on the list.

Chapter Seven
Antibiotics 101

When germs get past your skin and body fluids, your white blood cells and your antibodies, antibiotics are the last line of defense against germs.

What are antibiotics? Technically, **antibiotics** are substances that are **produced by germs** to kill other germs. Sounds weird doesn't it?

In the war of **germ against germ**, antibiotics are the weapons they hurl at each other. The mold *Penicillium notatum* doesn't want to share a Petri dish with bacteria, so it produces penicillin to kill the bacteria. Bacteria of the genus *Streptomyces* don't like to share their space with other bacteria, so they produce streptomycin to kill the other bacteria and so on. Bacitracin, cephalosporins and chloramphenicol are other examples of antibiotics.

In addition to the substances produced by molds and bacteria, there are substances produced in the lab that are used to treat infections. These substances are technically called **chemotherapeutic agents**, to indicate that germs didn't create them, but most people refer to them as antibiotics, too. We're going to use the word antibiotic for both antibiotics and chemotherapeutic agents as well. The chemotherapeutic agent you've heard about from the news, is ciprofloxin (aka **cipro**).

What makes a **good antibiotic**? You'd want an antibiotic to be something that **kills germs**. Drain cleaner and toilet bowl cleaner contain sodium hypochlorite, so they kill germs. They'd kill you too if you got a large dose of them in your body, so we have to be more specific about what makes a good antibiotic.

After germ-killing, the most important characteristic of a good antibiotic, is that it has what is called **selective toxicity**. Selective toxicity means that it kills the germs infecting you, but doesn't harm you. Ideally, the antibiotic wouldn't harm you at all, while it kills the infecting germs. Many good antibiotics fall short of that ideal. They're used because they're the best we have.

Some of the antibiotics used can cause some people to stop producing red blood cells. This phenomenon is called aplastic anemia. Other antibiotics can attack some people's nervous systems. There are a variety of bad side effects possible from some antibiotics. This is one reason why antibiotics shouldn't be used when it's known they can't help, such as using an antibacterial antibiotic to treat a cold or other virus infection.

Some people develop an allergic reaction to antibiotics. Penicillin is one of the common antibiotics that people may become allergic to.

Antibiotics are like other prescription drugs. They should be used with caution because they can cause problems for some people. If an antibiotic causes problems for everyone, it is usually pulled from the market by the Food and Drug Administration, unless there's a compelling reason not to.

How do antibiotics work? In general, antibiotics work by not allowing the invading germs to function normally.

Penicillin and some other antibiotics attack the **cell walls** of bacteria. When the bacterial cell walls are destroyed, the bacterial cells can "explode" like a balloon with too much air put into it. These antibiotics usually have a low toxicity for us, because our cells don't have cell walls. If this type of antibiotic causes a problem for us, it will probably be an allergic reaction.

Other antibiotics attack the **cell membrane** of the bacteria. This allows the cytoplasm and other cell-stuff to leak out of the cell. We have cell membranes, so these antibiotics can, on occasion, cause problems for us.

Some antibiotics interfere with **protein building** in the bacteria. If these antibiotics interfere with the **ribosomes** in the bacteria, they have low toxicity for us, because our ribosomes are different from the ribosomes in a bacterium. Unfortunately there are many other ways to keep a cell from making the necessary protein for enzymes and cell structures. Many of these can be a problem for us as well.

Still other antibiotics interfere with the **germ's genome**. Many **anti-virus antibiotics** work in this manner. We have DNA and RNA in our cells, so this is another kind of antibiotic that may be a problem for us.

An antibiotic will have different toxicity for different germs. There are many reasons for that. One of the most obvious reasons, is that when a germ produces an antibiotic, it isn't going to be killed by that antibiotic. Other reasons why germs differ in their sensitivity to an antibiotic, include resistance from previous exposures to the antibiotic and chemical differences in some of the cell structures.

An example of the chemical differences among bacteria, is the difference in the cell walls of the different kinds of bacteria. That difference is the first step in determining what bacterium is causing an infection. Bacteria, such as *Streptococcus pyogenes* (strep) and *Staphylococcus aureus* (staph) have cell walls composed primarily of a substance called **murein**. *Escherichia coli* and many other bacteria found in the intestines, have less murein and more fats and sugars making up their cell walls.

Penicillin attacks the murein, so that in general, it has less effect against the bacteria that have less murein in their cell walls. There are some notable exceptions to this rule, specifically the bacteria that cause gonorrhea and syphilis. Neither of these bacteria have cell walls that are mainly murein, yet penicillin works well against them.

If you've ever noticed that sometimes a **"Gram stain"** is ordered by someone on a television doctor show, it is to see what kind of cell wall the bacteria have. The Gram Stain procedure leaves the bacteria with mostly murein cell walls purple. The bacteria that have the fat and sugar layers in their cell walls are pink at the end of the Gram stain procedure. You can see the color under a microscope.

If you've ever wondered how doctors know what antibiotic to prescribe, there's a simple answer. They send a sample taken from your infection to the lab. The sample will mostly likely be a throat swab, some pus from a wound, urine, feces, blood, sputum or spinal fluid. The lab can do tests to see what antibiotic is the best to use.

You've probably heard from the news that some bacteria are resistant to many antibiotics. Misuse of antibiotics is the primary cause of antibiotic resistance. In the model for infection and treatment that follows, you'll see how populations of bacteria can be shifted so that most of the members of the population are antibiotic resistant.

A model for infection and antibiotic treatment

Imagine the person at the desk next to yours in class or at work has strep throat. The person coughs. You pick up 10 of the germs that were coughed into the air.

Let's keep this simple and assume that the *Streptococcus pyogenes* bacteria you picked up, double in number every 2 hours. I know you can do the math yourself, but you may not want to, so here's how the bacterial count looks over a 24-hour period. After:

2 hours, the 10 bacteria have become 20 bacteria

4 hours, the 20 bacteria have become 40 bacteria

6 hours, the 40 bacteria have become 80 bacteria

8 hours, the 80 bacteria have become 160 bacteria

10 hours, the 160 bacteria have become 320 bacteria

12 hours, the 320 bacteria have become 640 bacteria

14 hours, the 640 bacteria have become 1,280 bacteria

16 hours, the 1,280 bacteria have become 2,560 bacteria

18 hours, the 2,560 bacteria have become 5,120 bacteria

20 hours, the 5,120 bacteria have become 10,240 bacteria

22 hours, the 10,240 bacteria have become 20,480 bacteria

24 hours, the 20,480 bacteria have become 40,960 bacteria

and now you have a sore throat with a high fever and you are coughing and generally feeling miserable.

You go to the doctor. The doctor examines you and does a throat culture. You've probably had these before, where the doctor or an assistant holds your tongue out of the way with a tongue depressor, then runs a cotton swab over the back of your throat. The swab then goes to the lab for testing. Depending on the method used for testing, you may have the results immediately, or within a couple of days.

Your results come back with bad news. You have strep throat. The doctor gives you a prescription for penicillin and you have the prescription filled at the pharmacy. It calls for you to take several pills a day for about 10 days.

You go home and take your first penicillin tablet. Let's imagine that by now, you have 1 million *Streptococcus pyogenes* germs growing in your throat. Let's also imagine that these germs can be separated into groups of 100,000 germs, based on how easy they are to kill off with the penicillin. We'll give the weakest group a strength value of 1 and the strongest group a strength value of 10.

The penicillin tablet breaks up in your stomach and the penicillin is slowly absorbed into your bloodstream. Your blood transports the penicillin to your throat and kills all the germs with the strength value of 1. Now the germs you have left in your throat are those with a strength value of 2 or higher.

If the remaining germs in your throat double in number before your next penicillin tablet, you'll have 200,000 germs in each strength group of 2 or higher for a total count of 1,800,000 germs. The next tablet kills the germs with a strength level of 2, because you still have some penicillin in your blood from the first tablet, then you get more penicillin into your blood from the second tablet.

Again, the remaining germs double before your next penicillin tablet. This gives you 400,000 germs in each strength group of 3 or higher for a total of 3,200,000 germs. The germs are increasing in numbers and in their ability to resist being killed by the penicillin.

You keep taking your penicillin tablets for a couple of days and you begin to feel better, because the numbers of germs in your body are decreasing. The penicillin levels in your blood have built up to the point where they can kill off the tougher germs. Some of the toughest germs are getting old and tired, so they die too.

Let's say that you now have only 100,000 strep throat germs left in your body. The bad news is that these 100,000 germs are the "super-germs." They're tough enough to resist all the penicillin that you've taken so far. If you take *ALL*, the penicillin tablets in your prescription, the "super-germs" will eventually get old and tired of fighting the penicillin and die. When this happens, you will be cured of your infection.

If you stop taking your penicillin tablets, these super-germs will keep reproducing themselves, like you saw above in the 24-hour table. You'll get sick again, but this time the standard doses of penicillin may not kill them. You'll need a stronger antibiotic because these germs are resistant to the penicillin you took.

One of the main causes of antibiotic resistance is that people stop taking their antibiotic pills or liquid when they start to feel better, instead of taking the entire prescription. Another factor in helping germs develop resistance, is the use of antibiotics when they won't do any good, such as taking penicillin for a cold or other virus infection.

The antibacterial antibiotics kill the bacteria that are in your body. Some of the bacteria will survive the penicillin. If these are pathogenic bacteria, those that cause disease, they have successfully resisted the penicillin and are now among the antibiotic resistant bacteria.

So, now you know how bacteria multiply in your body, how they die off with antibiotic treatment and how they can become resistant to antibiotics. Yes, the example I gave is simple-minded, and the numbers are no where near being accurate, but it is a useful model for infection, antibiotic treatment and development of antibiotic resistance.

Antibiotics—a user's manual:

Many people complain that the antibiotics make them sick. Some people may find that the antibiotics are as unpleasant as the infection they are treating. Upset stomachs, diarrhea and for women, vaginal yeast infections, are a common result of taking antibiotics.

You can make your antibiotic treatment more pleasant in several ways.

1. First and most important of all, **talk to the pharmacist** about the antibiotic you'll be taking. Some antibiotics *must* be taken with food, others *can't* be taken with food. Make sure you know which kind of antibiotic you're taking and how to take it.

2. Ask your pharmacist if it's okay to drink a full glass of water with each antibiotic dose. If the pharmacist says it's okay to drink a full glass of water with your antibiotic, do it. The water will help dissolve the antibiotic pill and it will dilute the antibiotic in your stomach. This may reduce the stomach upset.

3. The next thing you can do for yourself is to **eat yogurt**, or some real sour cream or drink buttermilk, kefir or one of the new "biotic culture" drinks available in the dairy case. They have the good bacteria you need. If you have a problem with dairy foods, you can take acidophilus capsules available at any health food or vitamin store. Acidophilus capsules contain dehydrated *Lactobacillus acidophilus*, a friendly bacterium that helps prevent diarrhea and other intestinal upsets. Just make sure you're having your yogurt or buttermilk at times that are appropriate for the kind of antibiotic you're taking. You'll know what those times are, if you asked your pharmacist.

The reason you get the diarrhea and intestinal upsets from antibiotics, is that they kill the friendly bacteria in your body along with the bacteria that are causing your infection. When the bacteria in your intestines die, yeasts take over, causing diarrhea and sometimes, vaginal yeast infections.

The yogurt or acidophilus capsules put friendly bacteria back into your intestines. The simple explanation of why you need to put friendly bacteria back into your intestines is that the friendly bacteria take their share of the food, so the yeasts don't get it all and take over.

Chapter Eight
Bacterial Diseases

Bacteria cause many different diseases. As you'll see from reading through this chapter, one species of a bacterium can cause many different diseases, depending on what organ it infects or how it gets into your body.

Many of the **symptoms** of the diseases caused by bacteria are caused by **toxins** produced by bacteria. As you'll recall from the chapter on Toxins and Enzymes, there are two kinds of toxins: **exotoxins** and **endotoxins**. Exotoxins are excreted by the bacteria and are proteins. Endotoxins are part of the cell walls of the bacteria and are fat or if you want the official microbiologist's description: "The lipid A fraction of the gram negative bacterial cell wall."

Toxins are poisons. Some of the toxins are deadly. Others can make you so miserable for a few days that you might consider them deadly, too.

Some of the disease symptoms are actually caused by your immune system fighting off the toxins or germs. This knowledge won't make the symptoms any more pleasant.

Let's get acquainted with the **pathogenic bacteria** and the diseases they cause. You've heard of many of these germs already. Some you've met personally, others you've heard about from the news.

Actinomycetes

Actinomycetes are bacteria, but they resemble molds because they produce filaments. They are often found in **decayed teeth**. They may also be found in damaged tissue or cause **infections if your appendix ruptures**.

Bacillus anthracis

This bacterium produces an exotoxin that causes some of the tissue damage that is characteristic of the diseases associated with the organism. You can be vaccinated against *Bacillus anthracis*. *Bacillus anthracis* infections are treatable with antibiotics, but treatment must be started as soon as possible.

Bacillus anthracis produces spores that allow the organism to survive in conditions that actively growing bacteria could not survive, such as lack of water or nutrients. Spores of *Bacillus anthracis* are what were sent in letters in late 2001.

The diseases *Bacillus anthracis* causes are:

1. **cutaneous anthrax**. This disease has the following symptoms: skin lump that resembles spider bite, pus production, a skin ulcer or open sore that crusts over. Cutaneous anthrax results when spores get into an open wound. This is a form of direct contact spread of disease.

2. **pulmonary anthrax**. This disease has the following symptoms: flu-like symptoms, fever, headache, overall achiness, no runny nose, pneumonia, bloody fluid in your lungs, difficulty breathing, coma, and death if not treated in time. Pulmonary anthrax is caused by inhaling of spores. It is not a contagious disease, like flu. You can't get it from someone who has the disease.

Bacillus cereus

This organism is a relative of *Bacillus anthracis*. It also produces spores and exotoxin. There is no vaccination available for *Bacillus cereus*. Diseases caused by *Bacillus cereus* can be treated with antibiotics, though some of them usually aren't, because the illness is over in a day or two.

Diseases caused by *Bacillus cereus* are:

1. food poisoning. This disease has the following symptoms: nausea, vomiting, diarrhea, and abdominal cramps. Food poisoning is not treated with an antibiotic. The food poisoning is spread by meat or rice dishes that have the bacterium growing in them.

2. **keratitis**, an eye infection. Keratitis has the symptoms: avoidance of light, pain, tearing, and possible loss of vision. Keratitis can be treated with an antibiotic. Keratitis is caused by an eye injury

being contaminated by direct contact with the organism.

3. **systemic infections.** Systemic infections may be: heart, brain and bone infections or pneumonia. The systemic infections can be treated with antibiotics. The systemic infections are spread by contamination from I.V. drugs or medical devices.

Bartonella bacilliformis

This germ causes **oroya fever**, which is an **infectious anemia**. This disease is severe anemia. There will be skin lesions that come and go. This disease has a 40% mortality rate.

Oroya fever can be treated with antibiotics. The disease is spread by direct contact.

Bartonella heselae

This germ causes **cat-scratch disease**. Symptoms of cat-scratch disease are: papules or pustules—skin bumps with pus, low grade fever, headache, sore throat, conjunctivitis, and enlarged lymph nodes that may discharge pus.

Cat-scratch disease can be treated with antibiotics. Cat-scratch disease is spread by direct contact.

Bartonella quintana

Bacillary angiomatosis is caused by this germ. The disease has the symptoms: red papules with scale around them, erythema, open sores, and if internal organs are attacked, malaise, fever, abdominal pain, weight loss.

Bacillary angiomatosis can be treated with antibiotics. This disease is spread by direct contact.

Bordetella pertusis

This germ is the "P" of the "DPT" shot. You can and probably have been vaccinated against it. *Bordetella pertusis* causes **whooping cough**. This disease has the following symptoms: mild coughing initially, sneezing, violent coughing with "whooping" sound during inhalation, possibly vomiting, cyanosis (blueness), convulsions.

Whooping cough can be treated with an antibiotic. It is spread by air droplet.

Borrelia recurrentis

You probably guessed from the germ's name that it causes a disease that recurs. This germ causes a disease called **relapsing fever**. The symptoms are: sudden onset of fever, chills, weakness, headache, malaise, then you'll have a few days of feeling good. The cycle of feeling good and bad may repeat 3 to 10 times.

Relapsing fever is spread by tick bites. It can be treated with antibiotics.

Borrelia burgdorferi

This germ is another one that is spread by bites from ticks. You've probably heard of it—**Lyme disease**. The symptoms are: skin lesion at site of tick bite, flu-like symptoms of fever, chills, muscle pain and headaches. Other symptoms include joint pain, arthritis, neurological problems of meningitis or nerve palsy, heart disease with electrical conduction defects or inflammation of heart and surrounding membranes. Lyme disease can be treated with antibiotics.

Brucella species

This germ causes **brucellosis,** aka undulant fever. Symptoms of brucellosis: malaise, fever, weakness, aches, sweats, hepatitis and jaundice. Brucellosis is treatable with antibiotics. Brucellosis is spread by improperly pasteurized dairy food, animal urine, feces or contaminated milk. It is usually an occupational disease in people.

Calymmatobacterium granulomatis

This is a **sexually transmitted disease**. The name of the disease is **granuloma inguinale–genital ulcers or sores**. Granuloma inguinale can be treated with antibiotics. Granuloma inguinale is, of course, spread by sexual contact.

Campylobacter jejuni or Campylobacter coli

These germs are spread by fecal contamination. The disease caused by this germ is **diarrhea**. The symptoms are: diarrhea that may be bloody, fever, headache, abdominal pains. This diarrhea can be treated with antibiotics.

Campylobacter fetus

This germ causes **systemic infections**. This disease has the following symptoms: fever, pain, headache, shock, coma, and may be fatal. The infections can be treated with antibiotics. These systemic infections are spread through blood of immunocompromised patients.

Chlamydia species

These germs cause:

1. **trachoma or inclusion conjunctivitis, an eye disease** with the symptoms: tearing, mucous-pus discharge, conjunctival hyperemia—increased blood to the area, enlargement of eyelash follicles, blood vessels may grow into the cornea, causing scaring of the cornea, eyelid deformities and blindness. The disease can be treated with antibiotics. Trachoma or inclusion conjunctivitis is spread by direct contact.

2. **respiratory diseases** that infect infants or adults. Infants are usually infected during birth by mothers

who have the sexually transmitted *Chlamydia* infection. In adults the respiratory infection is usually mild. In infants, the infection is characterized by rapid breathing and no fever.

3. **sexually transmitted diseases.** These diseases may have no symptoms. They may cause small pustules in the groin region. The lymph nodes in the groin may also be involved and may have a discharge of blood and pus. If the lymph nodes are infected, the infection may spread to the rest of the body and cause headaches, nausea, vomiting, joint pain, fever and other symptoms of systemic disease. *Chlamydia* may cause **pelvic inflammatory disease** and may lead to sterility or ectopic pregnancies– meaning the fetus is in the fallopian tubes rather than in the uterus.

Clostridium botulinum

Clostridium botulinum is another spore and exotoxin producing bacterium. The exotoxin is a **neurotoxin** that attacks the victim's nervous system and causes **paralysis**. The paralysis is often **fatal**. The disease caused by this germ is **botulism**.

Vaccination against botulism is not generally available, though an experimental toxoid is used to immunize lab workers and people who work with animals. Antibiotic therapy for

Clostridium botulinum is not used. Antitoxin that is active against the exotoxin produced by *Clostridium botulinum* is used to treat botulism.

Botulism is an often fatal disease characterized by the symptoms of: visual disturbances, double vision, inability to swallow, slurred speech, no fever, respiratory paralysis, cardiac arrest, and death if not treated with antitoxin in time.

You develop botulism from ingesting or inhaling the toxin. There have been a few rare cases of wound botulism from puncture wounds being contaminated with the organism. Wound botulism develops like its relative, tetanus.

Home-canned vegetables are the most common source of botulism toxin. Spores of the *Clostridium botulinum* germinate in the canned food if it is not pressure cooked to kill the spores, during the canning process. Low acid foods, such as green beans are the most likely home canned foods to be contaminated with botulism.

If you can your own vegetables, **ALWAYS** boil the vegetables for at least 20 minutes before you eat them. More information on safe canning of foods can be obtained from your County Extension Service or from the manufacturers of canning jars. General recommendations on processing your food are usually included in the box with the canning jars.

There are several types of botulism toxin. They are referred by letters. Type A toxin is usually found in vegetables. Type E botulism is found more often in high protein foods, such as fish.

Clostridium tetani

This is another exotoxin and spore forming bacterium. It is usually found in the soil. You can be vaccinated against *Clostridium tetani* and the disease it causes, **tetanus**. The tetanus vaccination is against the toxin and is the "T" of the DPT shot. Tetanus can be treated with an antibiotic and antitoxin. Emergency treatment for tetanus is with antitoxin. You get tetanus when a wound, usually a puncture wound, is contaminated with spores.

Tetanus has the following symptoms: muscle contractions near the injury, jaw muscles contract so you can't open your mouth, which is why the disease is also known as '**lockjaw**'. All your muscles spasm, making it hard to breathe, which leads to death without treatment.

Clostridium perfringens

This spore and exotoxin producing organism has no vaccine available for preventing the diseases it causes. One of the diseases usually is not treated with antibiotics, because it goes away on its own in a few days.

The diseases caused by *Clostridium perfringens* are:

1. **food poisoning.** Like other kinds of food poisoning, you'll develop diarrhea, abdominal pain, nausea and often vomiting. The exotoxin produced by the

organism irritates your intestine to cause the unpleasant symptoms.

The food poisoning goes away when the toxin is flushed from your system. Fluid replacement and general supportive care are the only treatment.

A common source of this food poisoning is stuffing cooked inside a turkey. You can prevent this food poisoning by cooking the stuffing in a separate baking dish.

2. **gas gangrene**. This disease has the following symptoms: foul smelling discharge, tissue destruction, fever, blood destruction, toxemia—toxins in blood, shock, and death if not treated. The gas gangrene can be treated with an antibiotic. Surgical removal of the damaged tissue may also be required to contain the disease. It may be treated with antitoxin. Gas gangrene is caused by wounds becoming contaminated with spores.

Clostridium difficile

This is another spore and exotoxin producing organism. There is no vaccination against *Clostridium difficile*. The disease it causes, **pseudomembranous colitis**, can be treated with an antibiotic. This disease has the following symptoms: watery or bloody diarrhea, abdominal cramps and fever.

The pseudomembranous colitis results when antibiotics kill off normal flora (aka friendly bacteria) found in your intestinal tract. Eating yogurt or drinking buttermilk or kefir or taking acidophilus capsules may help prevent pseudomembranous colitis by replenishing the "friendly bacteria" in your intestines.

Corynebacterium diphtheriae

This organism produces an exotoxin, but not spores. You probably have been vaccinated against *Corynebacterium diphtheriae*. This vaccination is the "D" of the DPT shot. The disease *Corynebacterium diphtheriae* causes, is **diphtheria** and can be treated with an antibiotic.

Diphtheria has the symptoms: inflammation of respiratory tract, sore throat, fever, pseudomembrane in throat, suffocation, cardiac arrythmias, vision disturbance, speech disturbance, difficulty swallowing, and possibly death. The pseudomembrane that forms in your throat is caused by the exotoxin destroying the tissue in your throat. Diphtheria is spread by air droplet, like colds or flu.

Ehrlichia species

The disease caused by these germs is **ehrlickiosis**. This disease has the following symptoms: fever, chills, headache, muscle pain, nausea, vomiting, and rash. It can be treated with antibiotics. Ehrlickiosis is spread by bites.

Enterobacter aerogenes

This germ is found in your intestinal tract and in the soil. When this germ is found in water, it may indicate that the water was contaminated with raw sewage. The germ may be in the water from the soil. *Enterobacter aerogenes* causes **urinary tract infections** or **generalized sepsis**. Fecal contamination is usually the cause of the infections.

Entercoccus faecalis

As you can guess from the name, this organism is found in the intestinal tract.

This germ causes **nosocomial infections**–infections that are **acquired in the hospital**, spread on hands of personnel or by contaminated equipment. Nosocomial infections are characterized by pain at the site of infection, fever and sepsis. They can be treated with antibiotics.

Enterococci and Peptostreptococcus

These germs are found in your intestines. As long as they are in the intestines, they usually cause no harm. When they get into other organs or tissues of your body, they can cause serious infections. Fortunately, they are treatable with antibiotics.

Diseases caused by *Enterococci and Peptostreptococcus* are:

1. **urinary tract infections.** The symptoms are: painful urination, pus in urine, pain in your side. The germs get to your urinary tract from wiping following a bowel movement.

2. **wound infections**. These infections have the following symptoms: swelling, redness, pus production, the wound is painful to touch. These infections are caused by direct fecal contamination of the wounds. Proper hand washing will prevent many such infections.

3. **postpartum endometriosis**. This disease has the following symptoms: inflammation of uterine lining, abdominal pain, and may cause fever and vomiting. Treatment is with an antibiotic. Postpartum endometriosis is caused by contamination of the uterus during child birth.

4. **abdominal infection following ruptured appendix**. The symptoms are: abdominal pain, fever, and sepsis. Antibiotics treat the sepsis and infection following a ruptured appendix. The infection is from intestinal bacteria spilling out of the ruptured appendix. These infections can be fatal.

Erysipelothrix rhusiopathiae

This organism does not produce spores. It causes **erysipeloid**. The symptoms of erysipeloid are: infected cut, severe pain around injury, swelling, and a raised sore. Erysipeloid can be treated with an antibiotic. Erysipeloid is caused by a cut that becomes infected from fish or animal products.

Escherichia coli

This is a germ that has been in the news. You've heard about people becoming ill from contaminated beef that has not been properly cooked or from drinking contaminated fruit juices that were not pasteurized. There is no vaccination against *Escherichia coli*, better known as **E. coli**. Infections caused by this germ are treated with antibiotics.

Escherichia coli diseases are:

1. **diarrhea, hemorrhagic colitis, 'Montezuma's revenge'**. This disease has the following symptoms: diarrhea, blood in the stools, fever, abdominal pain. The diarrhea is spread by improperly cooked meat or from unpasteurized juices and dairy foods.

2. **wound infections**. This disease has the following symptoms: swelling, redness, pus, pain. Wound infections result from fecal contamination of wounds. A "rule of thumb" often used in hospitals is that when infected wounds are located between a person's knees and navel, look for *E. coli* in the wound.

3. **urinary tract infections**. These infections have the symptoms: frequent urination, painful urination, blood

or pus in the urine, pain in the side. *Escherichia coli* is found in your intestinal tract. It gets to your urinary tract following bowel movements.

4. **sepsis.** This disease has the following symptoms: pain, inflammation, pus production, fever, shock, and possibly death. *E. coli* sepsis is the result of fecal contamination.

5. **meningitis** with the symptoms: fever, headache, stiff neck, neural symptoms, coma and death if untreated. Meningitis results when germs get into your blood from other infections and spread to your brain.

Francisella tularensis

This germ is considered to be the **most contagious** germ in existence. There is some evidence that this germ can even get through healthy skin, which is a feat no other germs can accomplish. You can't be vaccinated against *Francisella tularensis*.

Francisella tularensis causes **tularemia**. This disease has the following symptoms: fever, malaise, headaches, pain in infected area and the lymph nodes near the site of infection, inflammation and necrosis of lymph nodes, bronchial inflammation and pneumonia. Tularemia can be treated with antibiotics.

Tularemia is spread by animal bites, contact with infected animals, and inhaling germs.

Fusospirochetal disease

This germ causes **trench mouth, tonsilitis, and gingivitis**. The symptoms are: inflammation, ulceration of the mouth, tonsils or gums.

These diseases can be treated with antibiotics. The diseases are spread by contact.

Gardnerella vaginalis or Gardnerella mobiluncus

These germs cause **vaginosus**. Vaginosus has more severe consequences for the fetus than for the mother. This disease causes preterm labor, premature birth, and may result in the death of the fetus.

Vaginosus can be treated with antibiotics. The disease is spread by sexual contact.

Haemophilus ducreyi

This germ causes **chancroid**, one of the lesser known **sexually transmitted diseases**. Chancroid appears as a ragged ulcer or sore on the genitals. It can be treated with antibiotics. Obviously, it is spread by sexual contact.

Haemophilus influenzae

This germ causes several different diseases. The diseases are treated with antibiotics. There is no vaccination. The diseases are:

1. **pneumonia** and other **upper respiratory infections**. These diseases have the following symptoms: coughing, sneezing, fever, congestion, flu, bronchitis or pneumonia symptoms. Pneumonia and other upper respiratory infections are spread by air droplet.

2. **tissue infections**, such as **cellulitis, septic arthritis**. The symptoms are: pain, inflammation, may have pus productions and tissue damage. These tissue infections result from wound infections spreading into deeper tissues.

Helicobacter pylori

This germ causes **gastric and duodenal ulcers**. Ulcer symptoms: burning sensation in the "stomach" or under the breastbone, nausea and diarrhea are possible.

Gastric and duodenal ulcers can be treated with antibiotics. The ulcers are spread by contamination.

Other Helicobacters

These germs cause **diarrhea**. The diarrhea can be treated with antibiotics. It is spread by contamination.

Klebsiella pneumonia

This is another germ that is from the intestinal tract. There are no vaccinations against this germ. Infections from it are treated with antibiotics.

Klebsiella pnuemonia diseases are:

1. **pneumonia**. This disease has the following symptoms: thick mucous filling air sacs, bleeding in the lungs, and tissue destruction. The pneumonia can be treated with antibiotics, but may also require surgery to remove the infected tissue. The pneumonia is caused by aspiration of germs.

2. **urinary tract infection**. This disease has the following symptoms: frequent urination, painful urination, blood or pus in the urine, and pain in the side. These infections are caused by fecal contamination.

Legionella pneumophilia

This germ got its name from the American Legion convention where it made its "grand entrance." It causes **Legionnaires' disease**. This disease has the following symptoms: sudden high fever, chills, malaise, non-productive cough, cyanosis from lack of oxygen, diarrhea, delirium, fluid in the lungs, and may be fatal.

Legionnaires' disease can be treated with antibiotics. Legionnaires' disease is spread by air droplets. It was first found in the water of the cooling units for the building that housed the convention where it got its name.

A milder form of the disease, Pontiac fever, has the symptoms: chills, fever, muscle pain, malaise, headache, dizziness, avoidance of light, stiff neck, confusion, and mild cough and sore throat. This disease also responds to antibiotic treatment and it spreads by air droplet.

Leptospira genera

The name for the disease caused by these germs is **leptospirosis**. Symptoms are: kidney and liver infection, jaundice, hemorrhage, meningitis with intense headache, stiff neck, muscle infection and eye lesions.

It can be treated with antibiotics. Leptospirosis is spread by contact with infected urine and feces.

Listeria monocytogenes

There are no vaccinations to protect against *Listeria monocytogenes* infections.

Antibiotics are effective against this organism and the diseases it causes. Those diseases are:

1. **fetal infections** that are contracted in the uterus. These infections are spread from mother to child. These infections develop into meningitis and often death.

2. **meningoencephalitis** in children and adults. This disease has the following symptoms: fever, headache and stiff neck. Meningoencepahilitis can develop into coma and death. These infections may start as a local infection caused by direct contact with

Listeria monocytogenes then spread to the brain through the body.

Mycobacterium tuberculosis

This is a germ that has had a significant impact on regional history. As you can guess from its name, the germ causes **tuberculosis**. Its impact on the history of Denver came in the late 1800's and early 1900's, when people believed that tuberculosis patients needed to go to places like Denver because that environment helped cure the disease. Because of that belief, TB patients flocked to Denver and were, literally, dying in the streets of Denver.

In 1899, a hospital was founded in Denver to treat tuberculosis patients. That hospital is now internationally known for advances in the treatment of lung diseases. This hospital is, of course, National Jewish Hospital. You can get more information on National Jewish Hospital and the lung diseases treated there, by going to:

http://nationaljewish.org/

The symptoms of tuberculosis are: inflammation, edema with fluid that resembles bacterial pneumonia, granuloma, tubercle or fibrous lesion. Tuberculosis is primarily a lung disease, but tubercles have been found in other organs, where they look like cancerous tumors on an X-ray.

Tuberculosis caused by some strains of the germ can be treated with antibiotics. *Mycobacterium tuberculosis* is one of the germs that has strains that have become **resistant to antibiotics**. Tuberculosis is spread by air droplet.

Mycobacterium avium complex

The disease that results from the *Mycobacterium avium complex* causes **organ dysfunction and bacteremia**. This disease has the symptoms of: bacteria in blood, organ dysfunction, fever, night sweats, abdominal pain, diarrhea, weight loss, lesions and infections in any part of the body. This disease primarily affects AIDS patients, and anyone whose immune system is compromised.

This disease can be treated with antibiotics. The disease is spread by air droplet.

Mycobacterium kansesii

This germ causes **pulmonary and systemic disease** that **mimics tuberculosis**. Because the disease mimics tuberculosis, it has the same symptoms: respiratory tract inflammation, edema, pneumonia-like symptoms, lesions, tubercle—a fibrous lesion.

It can be treated with antibiotics. Also like tuberculosis, this disease is spread by air droplet.

Mycobacterium scrofulaceum

This germ causes **chronic cervical lymphadenitis** with the symptoms of: inflammation of lymph nodes, granulomatous disease—granular lesions.

The disease can be treated with antibiotics. Chronic cervical lymphadenitis is spread by air droplet.

Mycobacterium marinum and Mycobacterium ulcerans

The disease caused by these germs is **skin lesions**. The lesions are called swimming pool granulomas because they are granular sores spread by infected water in lakes or swimming pools. The skin lesions can be treated with antibiotics.

Mycobacterium fortuitum

This germ can cause **systemic infections** that are characterized by sepsis, fever, shock, and possibly death.

The infection can be treated with antibiotics. This infection is spread by contaminated replacement heart valves that come from pigs.

Mycobacterium leprae

If you guessed from the name that this germ causes **leprosy**, you're right. The symptoms of leprosy are: macular lesions (bumpy sores) or red infiltrated nodules, neurologic symptoms of neuritis and parethesia, ulcers, bone destruction with resulting shortening of fingers and toes.

Leprosy can be treated with antibiotics. Leprosy is spread by contact.

Mycoplama pneumoniae

The pneumonia this germ causes is called **atypical pneumonia**. This disease has the following symptoms: lassitude (weakness), fever, headache, sore throat, chest pain, bloody sputum coughed up, systemic infection of nervous system, heart, joints and pancreas. Atypical pneumonia can be treated with antibiotics. It is spread by air droplet.

Mycoplasma hominis

This germ causes **genital and urinary tract infections**. The symptoms are: painful urination, frequent urination, may have blood or pus in urine, pain in the side, uterus and ovaries may become infected, and arthritis.

These infections can be treated with antibiotics. The infections are spread by direct contact.

Mycoplasma genitalium

This is another germ that causes **urinary tract infections** that are characterized by inflammation and painful urination.

These infections can be treated with antibiotics. The urinary tract infection is spread by direct contact or sexual contact.

Neisseria gonorrhoeae

If you look at the species name, you can guess what this germ causes. This germ does not produce spores or toxins. There is no vaccination against *Neisseria gonorrhoeae*. Infections can be treated with antibiotics.

It causes:

1. **gonorrhea** with the symptoms of: painful urination, mucous and pus discharge, may cause infertility in females if it reaches the fallopian tubes, skin sores with bloody or pus filled bumps, bacteria in blood, joint inflammation and pain.

   Gonorrhea is spread by sexual contact. Practicing safe sex by using a condom goes a long way toward preventing gonorrhea.

2. **ophthalmia neonatorum** which is a disease of newborns. This infection causes blindness.

   Ophthalmia neonatorum is spread from a mother infected with gonorrhea to the child during birth. This is one of the reasons most states require treating newborns with silver nitrate eye-drops. Other antibiotics may also be used to prevent the eye infection and blindness.

Neisseria meningitidis

This germ causes **meningitis**. This disease has the following symptoms: upper respiratory symptoms, like flu, high fever, sudden severe headache, vomiting, stiff neck, coma within a few hours, and death in many cases. This disease progresses so fast that when a spinal fluid sample is taken to the hospital lab, a technician drops everything to do an immediate gram stain and microscopic examination on the spinal fluid. A doctor or nurse is standing by, waiting for the results.

Fortunately, meningitis can be treated with antibiotics. This meningitis is spread by air droplet.

Other intestinal bacteria

"Anonymous" Germs from the intestinal tract cause infections that are treatable with antibiotics. The infections they cause are:

1. **urinary tract infections**— painful urination, frequent urination, blood or pus in urine, pain in the side.
2. **sepsis**— fever, shock, bacteria found in blood and organs. Sepsis is spread by germs getting into blood from other infections.
3. **wound infections**.

Pasteurella species

This germ causes **diseases associated with animal bites**. These diseases are characterized by: redness and swelling around the bite, pain in bite area, swollen lymph nodes, fever, respiratory infections, and bacteremia.

Diseases caused by Pasteurella species can be treated with antibiotics.

Propionibacterium acnes

You can probably guess one of the afflictions this organism causes, just by looking at its species name: "acnes" There is no vaccination against this organism. If there were, most of us would have demanded it as teenagers. Antibiotics treat infections from this organism. This is another exotoxin producing organism. Infections from *Propionibacterium acnes* are caused by direct contact with the organism. Skin or medical devices can be contaminated.

Propionibacterium acnes causes:

1. **acne**. No need to tell anyone the symptoms of acne. We all know them. Severe cases can be treated with antibiotics.
2. **infection of prosthetic heart valves**. The symptoms of the infection are fever, chest pain and generally feeling bad. These infections can lead to heart damage.
3. **infection of spinal shunts**. Spinal shunt infections have symptoms of: fever, headache, stiff neck and can develop into coma and death if not treated.

Proteus mirabilis or Proteus vulgaris

The source of these germs is fecal contamination. They are treatable with antibiotics. These germs cause:

1. **urinary tract infections** with following symptoms: painful urination, frequent urination, blood or pus in urine, pain in your side, kidney stones. These infections are treatable with antibiotics.

2. **pneumonia**. This disease has the following symptoms: chest pain, fever, bloody fluid in lungs. Aspirating the germs gets them into your lungs.

3. **bacteremia** with the symptoms: fever, sepsis, and shock. Bacteremia results from germs getting into blood from other infection sites.

Pseudomonas aeruginosa

Infections from this germ are unique because the **pus is blue-green** and **fluorescent**. The **blue-green pigment** this germ excretes is called **pyocyanin**. These infections can be treated with antibiotics.

Pseudomonas aeruginosa causes:

1. **burn or wound infections**. Symptoms are: redness, swelling, blue-green and fluorescent pus. Burn or wound infections are spread in the hospital or from contaminated water.

2. **meningitis** with the symptoms of: fever, brain inflammation, headache, stiff neck, coma, and death

if untreated. A meningitis patient is often infected by a spinal tap.

3. **urinary tract infection** with symptoms: frequent urination, painful urination, pus or blood in urine, pain in your side. These infections are often spread to urinary tract via catheters.

4. **necrotizing pneumonia**. This disease has the following symptoms: mucous forming in air sacs of lungs, cell destruction, fever, coughing up mucous, and the mucous may be blue-green. Necrotizing pneumonia is spread through respirators.

5. **sepsis** with symptoms: fever, bacteria in blood and organs, shock, coma, death, fluorescent pigment from red blood cell destruction. Sepsis is spread by germs getting into your blood from other infections.

6. **ecthyma gangrenosum**. This disease has the following symptoms: skin sores with red area around them, skin destruction causing bloody sores, blue-green pigment in pus, fluorescent pigment in pus. Ecthyma gangrenosum is spread by contamination of skin sores from germs.

Pseudomonads, Acinetobacters

These germs cause a variety of diseases, including: **pneumonia, eye infections, skin infections, urinary tract infections, septicemias, meningitis, endocarditis,**

nosocomial infections, bone and joint infections, gastrointestinal infections.

Symptoms of infections from these germs are: fever, inflammation, presence of bacteria, pain, other symptoms related to location of the infection. The infections are treated with antibiotics. The diseases are spread by contamination.

Rickettsia species

This family of germs causes several diseases:

1. **Typhus**. This is another disease that has had an impact on history. When armies have been infected with it, it has turned the tide of the battle to the other side. If you want to read the whole story, you can find it in a book titled: *Rats, Lice and History*, by Hans Zinsser.

   Typhus has the symptoms of: fever, rash, generalized body infection, prostration. Typhus can be treated with antibiotics.

   Typhus is spread by feces of fleas. The germs get into the body when the victim scratches flea bites and opens up the skin to admit the feces of the fleas.

2. **Rocky Mountain Spotted Fever**. Symptoms: pox-like rash of crusty bumps, fever, headaches, general body infection. You can be vaccinated against Rocky Mountain Spotted Fever. It can also be treated with

antibiotics. Rocky Mountain Spotted Fever is spread by tick bites.

3. **scrub typhus** with the symptoms of: fever, crusty rash, infection of brain or heart, inflammation. This disease can be treated with antibiotics. Scrub typhus is spread by arthropod bites.

4. **Q fever** whose symptoms are: flu or pneumonia symptoms, fever, headache, cough, fluid in lungs, hepatitis, brain inflammation, heart inflammation. Q fever can be treated with antibiotics. Q fever is spread by dust and aerosols.

Salmonella typhi

This germ causes **typhoid fever**. *Salmonella typhi* produces an endotoxin. Typhoid fever can be treated with antibiotics. The symptoms of typhoid fever are: fever, malaise, headache, constipation, slow heart rate, muscle pain, intestinal bleeding, possible intestinal perforation, hepatitis, and inflammation of abdominal organs. There may also be a bacteremia—bacteria in the blood from this infection. Typhoid fever is spread by fecal contamination of food or water.

Salmonella enteritidis

This germ causes a disease that goes by several names: **entercolitis**, **gastroenteritis**, or **food poisoning**. This disease has the following symptoms: nausea, headache, fever, vomiting, profuse diarrhea, and general misery.

Salmonella enteritidis produces an endotoxin. Food poisoning can be treated with antibiotics. It needs to be treated with fluid replacement to prevent dehydration.

Food poisoning is spread by contaminated food or water.

Serratia marcescens

This germ is treatable with antibiotics. There is no vaccination against the diseases it causes. *Serratia marcescens* causes:

1. **pneumonia** with the symptoms: bloody mucous in lungs, fever, chest pain. The pneumonia is spread by aspiration of germs.

2. **bacteremia**. This disease has the following symptoms: fever, shock, bacteria in blood. Bacteremia is spread by germs getting into blood from another infection.

3. **endocarditis** with the symptoms of: chest pain, fever, shock, inflammation of heart, valves and membranes around heart. Endocarditis is spread through blood stream from other infections.

Shigella dysenteriae

This germ causes **dysentery**. There is no vaccination against *Shigella dysenteriae*. The disease is treatable with antibiotics. The symptoms of dysentery are: watery diarrhea, blood and pus in stool, abdominal pain, and fever. The disease is spread by fecal contamination.

Spirilum minor

This germ causes a type of **rat bite fever**. This disease has the following symptoms: lesions at site of bite, skin rashes, swollen glands, relapsing or recurring fever.

Rat bite fever can be treated with antibiotics. This disease is spread by rat bites.

Staphylococcus aureus

This organism produces an exotoxin, but not spores. It has become highly resistant to most antibiotics. *Staphylococcus aureus* is the cause of many serious infections in hospitals.

Staphylococcus aureus causes many different diseases, ranging from skin infections to food poisoning to deadly toxic shock.

Note on *Staphylococcus aureus*. This organism is found in many people's nasal passages. Most of these people have no disease, they are simply carriers of the organism.

The diseases caused by *Staphylococcus aureus* are:

1. **local infections** often called **boils** or **carbuncles**. Boils are characterized by pimple-like bumps that are full of pus. Boils are spread by direct contact. They may require antibiotics.
2. **food poisoning**. This food poisoning can be distinguished from other types of food poisoning because there is no fever. Other symptoms of food poisoning are nausea, vomiting, and diarrhea.

This food poisoning cannot be treated with an antibiotic because the exotoxin excreted by the bacteria causes the symptoms. Because this illness is caused by the toxin, it is technically called a "**food intoxication**." The food poisoning is spread by contaminated food that is not kept refrigerated. Protein foods are common culprits.

3. **systemic infections**. The symptoms of systemic infections are: fever, pain, pus formed on infected organs that may develop into **toxic shock**. Systemic infections can be treated with an antibiotic. The systemic infections are spread by direct contact. You can get a *Staph* infection by touching contaminated dressings and bandages from a patient who has the infection. You can also spread it to other people if you don't wash your hands after touching those bandages.

Streptobacillus moniliformis

This germ causes **rat bite fever**. Rat bite fever has the symptoms: septic fever, blotchy petechial rashes–rashes that are caused by hemorrhages in capillaries, and polyarthritis. Rat bite fever can be treated with antibiotics.

Rat bite fever is caused by rat bites or by contaminated milk.

Streptococcus pyogenes

This is a particularly nasty germ. You've probably heard of "**flesh-eating bacteria**." Well, this is it, folks.

Streptococcus pyogenes doesn't produce spores. It does produce toxins and a variety of enzymes that allow it to eat its way through your tissues. There is no vaccination against *Streptococcus pyogenes*. The good news about this bacterium is that it is easily killed off by many antibiotics, like penicillin.

Streptococcus pyogenes produces Streptokinase. This is one of the enzymes that helps *Streptococcus pyogenes* eat its way through your body.

The diseases caused by *Streptococcus pyogenes* are:

1. **invasive infections**, such as **necrotizing fasciitis** (the medical term for the infection with flesh-eating bacteria) and **toxic shock syndrome**. These diseases have the following symptoms: redness around wound, tissue destruction, shock, bacteria in the bloodstream, respiratory failure, multiple organ failure, and death if not treated soon enough. Treatment is with antibiotics and may include surgery to remove the infected tissue.

    Necrotizing fasciitis and toxic shock syndrome are spread by air droplet or wound contamination. Toxic shock syndrome associated with the use of tampons is a direct contact spread.

2. **Scarlet Fever**. This disease has the following symptoms: fever, redness of entire body, shock, bacteria in the blood, respiratory failure, multiple organ failure, and death if not treated. Scarlet fever can be treated with an antibiotic. Scarlet fever is spread by air droplet.

3. **Erysipelus** is another disease caused by *Streptococcus pyogenes.* The symptoms are: swelling around wound, swelling enlarges as numbers of bacteria increase, fever, bacteria in blood. Erysipelus can be treated with an antibiotic. The disease is spread by wound contamination.

4. **puerperal fever AKA childbed fever**. This disease occurs when the uterus is infected during birth and the bacteria get into the mother's blood. The bacteria multiply as they circulate in the blood causing fever, shock, and may lead to death. Puerperal fever AKA childbed fever can be treated with an antibiotic.

    Puerperal fever was much more common before physician Joseph Lister convinced his colleagues to wash their hands between patients.

5. **sepsis or a generalized infections** are caused by the infection of surgical wounds or traumatic injuries. It is called surgical scarlet fever, because the wound and possibly the entire body will turn red. Other symptoms are fever, pain, shock and possibly death.

Sepsis can be treated with an antibiotic. It is spread by direct contact.

6. The best-known disease caused by *Streptococcus pyogenes* is **strep throat**. Strep throat is characterized by redness of the throat, enlarged lymph nodes in neck, ear ache, tonsilitis, pus discharge from mucus membranes, and fever. 20% of the cases have no symptoms. Strep throat is treated with antibiotics. It is spread by air droplet.

This disease has an unusually high carrier rate in the Rocky Mountain states of Wyoming and Colorado during the winter months. Strep throat was linked to rheumatic fever by research done at Warren Air Force Base in Cheyenne, Wyoming in 1949.

7. **Skin infections** caused by *Streptococcus pyogenes* are called **streptococcal pyoderma** aka **impetigo**. The symptoms are: blisters, open sores develop when blisters break, the sores ooze pus or crust over. It is treated with an antibiotic. The infection is spread by direct contact.

8. **Kidney infections** caused by *Streptococcus pyogenes* are called **glomerulonephritis**. This disease has the following symptoms: abdominal pain and painful urination. It may develop from skin infections and bacteria in the blood. Glomerulonephritis can be treated with an antibiotic.

It is spread by bacteria that get into the bloodstream from wounds.

9. *Streptococcus pyogenes* can attack your **heart** and cause **acute endocarditis**. This disease results when germs in your blood from another infection and travel to your heart. Bacteria settle on the heart valve or in the membranes inside the chambers of the heart. This infection causes destruction of heart valves, cardiac failure, death without antibiotic treatment or heart valve replacement. The infection can be treated with an antibiotic.

10. **neonatal infections**. This disease has the following symptoms: sepsis, or infection of an organ, meningitis, respiratory distress. These infections will occur within one month of birth if they occur. Neonatal infections can be treated with an antibiotic. The infant is infected during birth in many cases.

Post-Streptococcal diseases

These diseases are unique because they are caused when your **immune system** attacks your organs *after* **you've recovered** from a streptococcal infection. The short explanation of how this happens is that the *Streptoccocus pyogenes* tricks your immune system into thinking your organs are invading germs. Post-*Streptoccocal* Infections are:

1. **glomerulonephritis**. This disease has the following symptoms: blood in urine, protein in urine, edema, high blood pressure, urea nitrogen retention.

2. **rheumatic fever**. Again the person's immune system attacks an organ, in this case, the heart and valves. The symptoms are: malaise, fever, polyarthritis, and inflammation of all parts of the heart. There can be permanent heart damage.

Streptococcus viridans or Enterococcus

These germs are exotoxin producers that cause **subacute endocarditis**. Bacteria grow on the heart valves and membranes inside the chambers of the heart and cause heart murmur, fever, weakness, anemia, possible blood clots, enlarged spleen, and kidney lesions. Subacute endocarditis can be treated with an antibiotic.

Subacute endocarditis is spread by bacteria that get into blood from other infections and travel to the heart. Often the bacteria get into the blood from tooth extraction or other dental work that causes some gum bleeding. Defective heart valves are the main valves attacked by the organism.

Streptococcus pneumoniae aka Pneumococcus

This is the germ that causes many of the cases of **bacterial pneumonia**. It also causes other ailments. It produces an exotoxin. You can be vaccinated against *Streptococcus* *pneumoniae*. If you are in a high risk group, such as being over the age of 65, having heart disease or asthma, you should talk to your doctor about being vaccinated against *Streptococcus pneumoniae*. *Streptococcus pneumoniae* is treatable with antibiotics.

The diseases caused by *Streptococcus pneumoniae* are:

1. **bacterial pneumonia** with the symptoms of: sudden onset of fever, chills, chest pain, coughing up bloody or rust-colored sputum (the technical term for gooey mucous), bloody fluid and mucous fills air sacs in lungs. Bacterial pneumonia is spread by air droplet.

2. **ear and sinus infections**. The symptoms are: ear ache, redness of ears, fever, sinus pain and headaches from the sinus pain. Ear and sinus infections are caused by aspiration of germs that got into your throat and nose from droplets.

3. **meningitis**. This disease has the following symptoms: headache, fever, stiff neck, neural symptoms, coma. Meningitis occurs when germs spread through body from pnuemonia or other infection.

Treponema pallidum

The disease caused by this germ is one of the best-known **sexually transmitted diseases—syphilis**. There are several forms of syphilis:

1. **primary**. This form is the initial infection that has the symptoms: papule or lump, genital ulcer or sore with a hard base called a hard chancre. This form can be treated with antibiotics.

2. **secondary**. This form follows the primary form if it is left untreated. Symptoms are: red maculopapular rash, pale bumps in anal or genital area, bumps in armpit or mouth, may develop into meningitis, eye disease called chorioretinitis, liver inflammation, better known as hepatitis, kidney disease, inflammation of the membrane covering the bones.

3. **tertiary**. This form develops from the secondary form when it is untreated. The symptoms are: granular lesions of skin, bone and liver, central nervous system degeneration, cardiovascular lesions, inflammation of the aorta, aortic aneurysm and heart valve insufficiency.

4. **congenital syphilis**. This form is transmitted from an infected mother to the fetus. This form causes miscarriage and still birth, or keratitis, saddle-nose, bone inflammation and nervous system abnormalities in surviving infants.

Ureaplamsa urealyticum

This is one more germ that causes **urinary tract or genital infections**. This disease has the primary symptom of inflammation or the urinary or genital tract. If a child is infected at birth, the germ may cause **lung infections**.

The infections can be treated with antibiotics. They are spread by direct contact or sexual contact.

Vibrio cholera

This germ produces an endotoxin. Vibrio cholera can be treated with an antibiotic. You can be vaccinated against *Vibrio cholera*. Cholera is spread by contact with infected feces.

The disease caused by *Vibrio cholera* is **cholera**. This disease has the following symptoms: nausea, abdominal cramps, profuse, watery diarrhea with bits of mucous, termed "rice-water" stools in many text books, dehydration, circulatory collapse if patient not re-hydrated. Replace electrolytes and fluids with electrolyte drinks available in the pharmacy, sports drinks or water with sugar added to help your intestines absorb the water.

Other vibrios

These germs cause gastroenteritis, wound infections, cellulitis and bacteremia. There is no vaccination against these germs. Antibiotics treat the infections from them.

Diseases are:

1. **gastroenteritis** with the symptoms: nausea, vomiting, fever, abdominal cramps, watery or bloody diarrhea and dehydration. Fluid replacement is

essential. The disease is spread by fecal contamination.

2. **wound infections, cellulitis, bacteremia.** These infections have the following symptoms: fever, edema, inflammation and pus production. These infections are caused by fecal contamination or spread from other infections.

Yersinia pestis

This germ has had a huge impact on human history. In the Middle Ages it killed millions of people throughout Europe. You can be vaccinated against *Yersinia pestis* though it isn't usually done for most of us. This germ produces an exotoxin.

The diseases caused by *Yersinia pestis* are:

1. **bubonic plague aka black death**. This disease has the following symptoms: high fever, enlarged and painful lymph nodes in the groin and armpits—called buboes, vomiting, diarrhea, blood clotting, altered mental state, kidney and heart failure, pneumonia and meningitis symptoms prior to death. Bubonic plague can be treated with antibiotics. Bubonic plague is spread by flea bites.

2. **pneumonic plague**. This disease has the following symptoms: high fever, enlarged lymph nodes, bleeding into lungs, consolidation—lungs filling with fluid, low blood pressure, blood clotting, altered

mental state, sepsis, heart and kidney failure, and death. Pneumonic plague can be treated with antibiotics. It is spread by air droplet like colds or flu. *Yersinia pestis* is believed to be a biological weapon.

Yersinia enterolitica and Yersina pseudotuburculosis

These germs cause **generalized infections**. This disease has the following symptoms: fever, abdominal pain, watery or bloody diarrhea, joint pain and arthritis, pneumonia, meningitis and sepsis. The infection generally self-limiting—meaning it goes away on its own.

Yersinia enterolitica and Yersina pseudotuburculosis infections can be treated with antibiotics. *Yersinia enterolitica and Yersina pseudotuburculosis* produce an enterotoxin.

The disease is spread by food or water that are contaminated with animal feces or contaminated with the toxin produced by the germs.

Chapter Nine
Virus Diseases

As you will recall from the chapter on the germs and the scientists who study them, **viruses are specific to their hosts**. Animal viruses only attack animals. Plant viruses only attack plants. Animal viruses may have only one or two possible animals as their hosts, such as the virus that causes canine distemper or they may attack many different animals, such as rabies virus. Viruses need hosts that have the enzymes that allow the viruses to reproduce themselves.

Types of vaccines for virus diseases:

There are vaccinations against some of the viruses. There are 3 types of vaccines against virus diseases. These are **killed-virus**, **attenuated live-virus** and **live-virus**. To explain the difference, we're going to consider the following analogies.

Imagine that a rattlesnake represents the virus. When you see a rattlesnake, you recognize it as a snake, just as your immune system "sees" a virus and recognizes it as an invading germ.

When you see a dead rattlesnake, you still know it's a rattlesnake. When your immune system "sees" a **killed-virus**, it recognizes the virus. Neither the dead rattlesnake nor the dead virus can hurt you but you and your immune system see them, recognize them and react. The original **Salk Polio vaccine** given in the 1950's and early 1960's was a killed virus vaccine.

An **attenuated live-virus** vaccine is like a rattlesnake whose fangs have been removed. The snake is still poisonous, but it can't get the poison into you. The attenuated-live virus lacks the "fangs" that would allow it to make you sick, but your immune system still recognizes it and makes antibodies against it. The **Sabin Polio vaccine** that came into use in the mid-1960's is an attenuated live-virus vaccine.

Finally, there is **live-virus vaccine**. This is like the rattlesnake that still has its fangs and the ability to produce venom. Live-virus vaccines actually **infect you**. The live-virus vaccine you've been hearing about on the news—**smallpox**, causes an infection. Because you're infected with the virus from the smallpox vaccine, you're also contagious from being vaccinated.

Because the smallpox vaccine actually infects you, it is more dangerous than the other two kinds of vaccines. Infections are dangerous. Some people recover from an infection with no problems, others become extremely ill from the infection and a few die from the infection. This is why there has been so much debate about whether or not to give the smallpox vaccinations to everyone as a precaution against a possible bioterrorism threat.

You may be wondering why most people can be vaccinated with live-virus smallpox vaccine without becoming critically ill with smallpox. The answer is that the virus in the vaccine isn't the smallpox virus. It's **cowpox virus**, so if you're vaccinated against smallpox, you're being infected with live cowpox virus.

Cowpox virus is used as a vaccine against smallpox, because a physician, Edward Jenner, noticed that people who milked the cows often developed sores on their hands, but didn't get sick during a smallpox epidemic. In the 1790's, Jenner hypothesized that the infection with the cowpox protected the people against smallpox. He vaccinated his children with cowpox and exposed them to smallpox. Jenner's children didn't get smallpox.

An interesting historical note: the term 'vaccination' comes from the medical name of the cowpox virus—Vaccinia.

Primary treatment for virus diseases:

The primary treatment for virus infections is what is termed "**supportive care**" which is the medical term for treating the symptoms and preventing dehydration. You know all about supportive care. It's how you treat your colds.

There are only a **few antibiotics** that are active against viruses. These antibiotics are recent medical discoveries. In contrast, the best-known antibiotic that is active against bacteria, penicillin, was discovered in 1928 and went into mass production during World War II.

The Viruses:

Let's look at the viruses that cause human diseases. Most of the virus groups are based on the type of **nucleic acid the virus has**. You'll recall that a virus has **only** DNA or RNA, never both. One notable virus group, **Arboviruses**, is based on how the **viruses spread.** These viruses are spread by insects whose scientific designation is arthropods.

AIDS and lentiviruses

AIDS and lentiviruses cause **Acquired Immune Deficiency Syndrome** which attacks the protective white blood cells in your body. Because your protective white blood cells are attacked by the AIDS virus, you are vulnerable to many diseases. These viruses don't have specific symptoms, other than a weakened immune system. If these viruses kill you, it's because you become infected with something that your immune system can't fight off.

If you need to disinfect something that might have been contaminated with blood from someone who has AIDS or is HIV positive, **these viruses can be killed by:**

a 10% mix of household bleach and water

50% ethyl alcohol (beverage alcohol)

rubbing alcohol (70% isopropyl alcohol)

.5% and .3% hydrogen peroxide

Any of the chemicals listed above can be used to disinfect items that have been contaminated with blood or other body fluids from an AIDS patient. Disinfection requires at least 10-15 minutes of soaking in the disinfectants. Medical and dental instruments should be sterilized by steam autoclaving.

AIDS is spread by sexual contact, blood and blood products, and mother to child across the placenta. There is no evidence of spread in saliva or from insect bites. Antibiotic treatment is available but not consistently effective. Experimental vaccines are being tested.

Arboviruses

Arboviruses are grouped by how they are spread. The name **"ARBO" is from <u>AR</u>thropod <u>BO</u>rne**. Common arthropods that spread viruses are ticks, fleas, lice, mosquitoes. The diseases that are caused by arboviruses are:

1. **Dengue fever** with the symptoms: fever, muscle and joint pain, swollen lymph nodes, may develop into hemorrhagic fever that has a high mortality. No vaccinations or antibiotics are available for this disease.

2. **West Nile fever** that has the symptoms: mild fever, swollen glands, rash. West Nile fever may develop into fatal encephalitis. This disease is spread by mosquito bites as you've probably heard from the news.

3. **Sandfly fever** with the symptoms: fever, headache, malaise, nausea, stiff neck and back, eye pain, avoidance of light, abdominal pain. As you might have guessed from the name of the disease, sandfly bites spread it.

4. **Rodent-borne hemorrhagic fevers** have the symptoms of headache, nausea, vomiting, may develop into kidney problems or fluid in the lungs which can be fatal. Hantaan virus is an example of these diseases. These diseases are spread to humans by rodents and rodent droppings.

5. **Lassa fever** whose symptoms are: high fever, muscle pain, mouth sores, skin rashes with hemorrhages, pneumonia, heart and kidney damage. This is another disease spread by rodents and rodent droppings. Antibiotic treatment is available.

6. **Encephalitis diseases**, such as Western, Eastern, Venezuelan equine encephalitis. West Nile fever falls into this category of disease, too. These diseases have the symptoms: severe headaches, fever, vomiting, lethargy, convulsions. Encephalitis diseases are spread by mosquito bites.

7. **Marburg and Ebola** cause diseases that have an extremely high mortality rate. The symptoms are: fever, headache, sore throat, muscle pain, abdominal pain, vomiting and diarrhea, rash and internal and

external bleeding. These viruses have been "immortalized" in some medical or technical thriller novels and films.

These viruses are **killed by UV light, heating for 30 minutes or by household bleach**. It is unknown how these diseases spread. Rodents or animal bites are the suspected modes of spread. There is no antibiotic treatment or vaccination available. Supportive care is the only treatment.

8. **lymphocytic choriomeningitis** with the symptoms: fever, malaise, muscle pain, weakness, sore throat, cough, acute meningitis. This disease is spread by rodents and rodent droppings.

9. **yellow fever** a hemorrhagic fever. The symptoms are jaundice—yellowing of skin, protein in urine, fever and chills, head and backache, black vomit from coagulated or partially digested blood. There is a high mortality rate. This disease is spread by mosquito bites. Vaccination is available for yellow fever.

Adenovirus

Adenovirus causes several diseases:

1. **respiratory diseases**, including **viral pneumonia** with the symptoms: cough, nasal congestion, runny nose, fever and chills, malaise, headache, muscle

aches, sore throat. Adenovirus respiratory diseases are spread by air droplet. There is no antibiotic or vaccination available.

2. **conjunctivitis** is an infection of the membrane over your eye. It has the symptoms: redness, tears, irritation, sensitivity to light. This is spread by air droplet or direct contact with an eye. Again, no antibiotics or vaccinations are available.

3. **gastroenteritis** with the symptoms: diarrhea, abdominal cramping. This disease is spread by air droplet. No antibiotics or vaccinations available. Treatment is supportive care and replacing fluids.

4. **hemorrhagic cystitis**, a bladder infection with the symptoms: painful urination, frequent urination, blood in urine. This disease is spread by direct contact. Adenovirus is **associated with cancers**.

NOTE about medical use of Adenovirus: It is used in gene therapy to carry the gene into cells.

Calciviruses

Calciviruses cause **epidemic gastroenteritis** with the symptoms: diarrhea, nausea, vomiting, low-grade fever, abdominal cramps, headache, malaise. This virus is spread in food, particularly shellfish, and water. No antibiotics or vaccinations are available.

Coltivirus is also grouped as an Arbo virus

Coltivirus causes **Colorado tick fever**. The symptoms of Colorado tick fever are fever and chills, along with muscle and joint pain, backache, eye pain, nausea and vomiting. This disease is spread by tick bites. Antibiotic treatment is not used. Vaccination is available.

Coronaviruses

Coronaviruses cause the **common cold.** They also cause **gastroenteritis** whose symptoms are vomiting and diarrhea. These diseases are spread by air droplet.

Herpesvirus Family

This virus family includes:

A. Herpesvirus which causes lifelong infections that come and go. When there is no active infection, such as a cold sore, the virus is latent or "hibernating." Viruses multiply at the site of the infection then invade the nerves and migrate away from the infection site. Herpesviruses are spread by direct contact. Herpesvirus diseases are:

> 1. **mouth and throat diseases** with the symptoms: sores in mouth that may be fluid filled sacs or open sores, fever, sore throat, swollen glands under the ears. If your gums are involved, they'll be swollen. Herpesviruses may also cause tonsilitis. Cold sores

are the best-known Herpesvirus infection. Some antibiotics are available.

2. **keratoconjunctivitis** with the symptoms: sores or ulcers on eyelid or cornea. The cornea becomes opaque causing blindness. This disease is spread by direct contact with body fluids of a person who is excreting the virus. Antibiotic treatment is available. Vaccinations are not available.

3. **genital herpes** has the symptoms: sores on penis in men, sores or lesions on cervix, vulva, vagina or perineum in women, fever, painful urination, swollen lymph nodes in the groin. This disease is spread by direct contact. Some antibiotic treatment is available. No vaccination is available.

4. **localized skin infections** with the symptoms: lesions or sores on the fingers or the body. Lesions are usually on the fingers of dentists or hospital personnel. Wrestlers may get the lesions on areas of their bodies other than their fingers. The spread is by direct contact. Antibiotic treatment is available. Vaccination is not available.

5. **severe, life-threatening skin infections** that usually occur only in people who have other chronic skin problems, such as eczema or people who are burn patients. Symptoms are vesicles or fluid-filled bumps all over the body and high fever. This disease is

spread by direct contact. Antibiotic treatment is available. No vaccination is available.

6. **encephalitis** with the symptoms: fever, headache, stiff neck, coma, possibly death. This disease is spread by direct contact with the virus. Antibiotic treatment is available.

7. **neonatal herpes** has the symptoms: sores or lesions on eyes, skin or mouth, encephalitis, multiple organ disease. This disease has approximately 50% mortality. Neonatal herpes is spread during birth. The infant is infected by an infected mother. Caesarean sections are often used to deliver babies of mothers infected with herpes. Antibiotic treatment is not available for the babies.

8. **disseminated infections** that usually strike transplant or AIDS patients, because their immune systems are suppressed. These infections cause lesions of the respiratory tract, esophagus and mucous membranes of the intestine. These infections may be fatal in malnourished children. Spread is by direct contact. Neither antibiotic treatment nor vaccination is available.

B. Varicella-Zoster Virus

Varicella-Zoster Virus causes:

1. **chicken pox** with the symptoms: vesicles on skin and mucous membranes, fever, severe itching that can cause scarring if sores are scratched, sores crust over in later stages. Chicken pox is spread by air droplet. Antibiotic treatment is available. It is also treated with gamma-globulins. Vaccination is available, and required for school-age children in many states.

   Never give aspirin to children or teens with chicken pox *without consulting your family physician first, because of the risk of them developing **Reye's syndrome**.*

2. **shingles** with the symptoms: pustules that crust over, pain that lasts long after the skin lesions heal. This is the same virus that causes chicken pox. The disease is spread by air droplet. Antibiotic treatment is available or it may be treated with gamma-globulins. Vaccination is the same as for chicken pox.

C. Cytomegalovirus

Cytomegalovirus is another virus that causes several different diseases:

1. **infectious mononucleosis** that has the symptoms: fever lasting for long time, malaise, muscle pain, swollen neck glands, severe fatigue and weakness. This disease is spread by direct contact with body

fluids of an infected person. Antibiotic treatment is available, but vaccination isn't available.

2. **pneumonia and disseminated disease** in immune compromised patients with the symptoms: gastroenteritis, chorioretinitis that leads to blindness. Again, the spread is by direct contact. Antibiotic treatment is available.

3. **congenital infections** that are contracted by the fetus while it is in the uterus. The symptoms are severe neurological problems, such as hearing loss, vision problems or mental retardation, fetal death is common. No antibiotic treatment or vaccination is available.

D. Epstein-Barr Virus

Epstein-Barr Virus causes several diseases:

1. **infectious mononucleosis** with the symptoms: headache, malaise, fatigue, fever that lasts approximately 10 days, sore throat, illness lasts 2-4 weeks. This disease is spread by direct contact with an infected person's body fluids, such as saliva. Antibiotic treatment is available. No vaccination is available.

2. **chronic infections** with the symptoms: pneumonitis, hepatitis, blood cell abnormalities, prolonged and

relapsing illness. This disease is spread by direct contact.

3. Epstein-Barr Virus is **found in some tumors**.

4. **Oral hairy leukoplakia** is a wart-like lesion that occurs on the tongue. This disease generally only strikes immune compromised patients. This disease is spread by direct contact.

E. Other Herpes Viruses

Other Herpes Viruses are found in tumors and associated with rashes and fevers. They are usually spread by direct contact.

Orbiviruses

Orbiviruses cause **mild fevers**. Spread is air droplet.

Orthomyxoviruses–influenza

Orthomyxoviruses or influenza viruses cause **Flu or influenza**. The symptoms are: chills and fever, headache, muscle aches, respiratory symptoms of sneezing, coughing, runny nose, ear infections in children. Bacterial infections, such as bacterial pneumonia, may follow flu in elderly or debilitated patients. The bacterial infections that follow flu are the cause of most of the flu deaths during an epidemic. There can be millions of deaths during worldwide epidemics (called pandemics). A pandemic that began in 1917 was known as

"Spanish Flu." It killed more than 25 million people throughout the world.

This disease is spread by air droplet. Antibiotic treatment is available. Vaccination available and recommended for people at high risk of developing flu, such as people who have heart disease or chronic lung problems like asthma.

Orthomyxovirus also causes **Reye's Syndrome.** Reye's syndrome is an acute brain infection with liver damage. It is associated with use of aspirin treatment in teens and children with flu. **Never use aspirin to treat the fever with flu in children and teens** without consulting your family physician first. Consult your physician for advice on alternate fever reducing drugs.

Respiratory synctial virus is another orthomyxovirus. It causes **lung infections** and lower respiratory tract infections, such as pneumonia. The symptoms are inflammation and fluid swelling that blocks the airway, wheezing. The virus may also cause ear infections. Treat this disease with oxygen and remove the secretions from the throat. This disease is spread by air droplet. Antibiotic treatment is available.

Paramyxoviruses

Paramyxoviruses cause several diseases:

1. **common cold, croup or bronchitis** with the symptoms: fever, runny nose, cough, headache and body aches. If the person has croup the cough may

be a "barking" type. This disease is spread by air droplet.

2. **mumps** has the symptoms: enlargement of glands in neck and fever. It may spread to ovaries or testicles, and cause encephalitis. Testicular damage may result, particularly when adults are infected. This disease is spread by air droplet. Vaccination available and required for school age children in some states.

3. **rubella or 3-day measles** has the symptoms: malaise, low-grade fever, red rash on face, trunk and extremities. Rubella is often mistaken for other infections that are characterized by fever and red rash on the body. A blood test for presence of the antibodies for rubella is required for conclusive diagnosis of rubella. This disease is spread by air droplet. Vaccination is available and required for school age children in some states.

4. **congenital rubella** results when the mother is infected during pregnancy. Congenital rubella leads to congenital heart disease in the child, total or partial blindness or deafness, slow growth of the child, enlarged liver and spleen, anemia, bone or brain inflammation, learning difficulties, problems with balance and muscle control. The symptoms may be obvious at birth or may develop later, including brain

deterioration in late teens. Congenital rubella has a high mortality rate. This disease is spread by mother to child. Vaccination is available for the mother.

Parvovirus

Parvovirus causes **chronic anemia** in humans that has flu-like symptoms in the early stage, fever and chills, malaise, itching, joint pain like rheumatoid arthritis, skin rashes in children, aplastic anemia—no red blood cell production, may be fatal.

This disease is spread by air droplets, contact with blood or mother to fetus. There is no antibiotic treatment or vaccination for humans.

This virus causes the disease "**parvo**" in dogs and cats. Pets can be vaccinated against the virus.

Picorna Virus group includes:

A. Cocksackie Virus. In general, Cocksackie virus is spread by throat swabbings, feces and flies. It causes several diseases:

1. **aseptic meningitis** that has the symptoms: fever, malaise, headache, nausea, abdominal pain, muscle weakness and paralysis that is usually reversible. Spread by air droplet.

2. **herpangia** that has the symptoms: severe sore throat, fever, vesicles on palate, throat, tonsils,

tongue. This disease is self-limiting, meaning it goes away by itself. This disease is spread by air droplets.

3. **hand-foot and mouth** with the symptoms: ulcers in mouth and throat, rash with fluid filled bumps, crusting as the vesicles heal. Viruses are shed in feces, saliva and the fluid from the vesicles. This disease is spread by direct contact.

4. **pleurodynia** that has the symptoms: loss of appetite, malaise, headaches, followed by stabbing chest pain and abdominal pain. This disease is spread by air droplet.

5. **myocarditis or pericarditis** is inflammation of heart or of the membranes around the heart. Myocarditis and pericarditis may cause permanent heart damage. The symptoms are chest pain and fever. The infections may be fatal in newborns. Alcohol, steroids, pregnancy and exercise make the symptoms worse. This disease is spread by direct contact or inhalation.

6. **acute hemorrhagic conjunctivitis** is an infection of the membranes of the eyes. The symptoms are red eyes and tears. This disease is spread by direct contact.

7. **common cold** with the usual cold symptoms. The common cold is spread by droplet. Treatment is to treat the symptoms and rest.

8. **diarrhea** that is spread by droplet.

9. **generalized disease in infants**. This is multiple organ attack by the viruses. This type of infection is often fatal. The infection spreads mother to child across the placenta.

10. **related to insulin dependent diabetes** aka type I.

11. **related to chronic fatigue syndrome**. Symptoms are exactly what the name suggests, being extremely fatigued for long periods of time.

B. Echovirus

Echovirus gets the name "echo" from a description of it: **E**nteric **C**ytopathic **H**uman **O**rphan. These viruses cause: **aseptic meningitis, encephalitis, fevers** with and without rashes, and the **common cold**. Symptoms of the diseases that are caused by this virus, are the same as symptoms of same diseases caused by other viruses. This disease is spread by air droplet and other methods. No antibiotic treatment or vaccination is available.

C. Rhinoviruses

Rhinoviruses cause several diseases:

1. foot and mouth disease of cattle with the symptoms: fever, salivation, fluid filled sores around mouth and feet. The sores rupture and spread virus. This disease is spread by contact with fluids from sores

and may be spread by migrating birds. There is no antibiotic treatment for foot and mouth disease of cattle. Vaccination is available, but the immunity is short-lived.

2. **the common cold** with the symptoms: fever, sneezing, coughing, runny nose, sore throat, body aches, headache. This disease is spread by air droplet, usually from droplets on hand or other object. No antibiotics or vaccinations available. Some people believe that various home remedies and zinc throat lozenges or zinc nose sprays help shorten the duration of the common cold. Each person must judge the value of these treatments for himself.

D. Hepatitis Virus

The hepatitis Virus group causes **infectious hepatitis** types A, B, C, D, E. The symptoms are fever, nausea, vomiting, jaundice from inflammation of the liver, skin rash, arthritis in many joints, damage to blood vessels or kidney. Type A hepatitis virus is often spread in food. All types of hepatitis spread by contact with blood, urine and feces. Antibiotic treatment is available. Types A and B are often treated with gamma-globulins. Interferon, a host-cell substance that interferes with viruses, is also used to treat the infections. Vaccinations are available for types A and B.

E. Enteroviruses

Enteroviruses cause **poliomyelitis**, which is also called **infantile paralysis or polio**. The disease has the symptoms: fever, malaise, drowsiness, headaches and back and neck pain, nausea, vomiting and constipation, sore throat, muscle weakness, lack of coordination and muscle spasms in non-paralyzed muscles. These viruses may cause muscle atrophy in aging patients who had polio as children.

Historical note: President Franklin Delano Roosevelt had polio as an adult and was paralyzed by it. He suffered from paralysis while serving as President of the United States.

This disease is spread by ingestion of viruses. Milk pasteurization inactivates enteroviruses. No antibiotic treatment available. Vaccination is available. There are two types: Salk vaccine (killed virus) and Sabin (attenuated live-virus).

Until the mid-1960's, polio was a major health threat. Some people developed forms of polio that paralyzed the muscles used for breathing. They were treated in devices called "iron-lungs" that helped them breathe without having to have a respirator tube down their throats. More information about polio and many other health threats is available on the CDC's web site at: www.cdc.gov. Click on the Health A-Z link and select polio from the list to read more about it.

Poxvirus family

Some of the viruses in the poxvirus family are in the news. The members of the poxvirus family are:

A. Vaccinia

Vaccinia causes **cowpox** with the symptoms: fluid-filled lesions or sores. These sores are usually on the hands of the people who work with the animals that are infected with cowpox. The sores crust over in later stages. This disease was common among people who worked with cattle. This disease is spread by direct contact with the pox sores on infected animals.

B. Variola

Variola causes **smallpox**. There is a **2 week incubation.** The initial symptoms are malaise and fever that lasts about 5 days, then the skin sores appear. There will also be sores in the mouth and respiratory tract. The sores are called papules. Papules are bumps for the first few days, then they become filled with fluid. Later the papules become filled with pus, break open and crust over. The crusts fall off in a few weeks. The fever usually drops after the rash appears, rashes may be hemorrhagic. The greater the number of bumps, the more severe the disease is.

The 2 week incubation of smallpox is important to note. This is why you've heard on the news that you can be immunized against smallpox even after you've been exposed to the disease.

To see a picture of the papules of smallpox, go to: http://www.cdc.gov, choose A-Z topics and find smallpox.

Smallpox was declared eradicated in 1979. A few labs have stocks of the viruses. Smallpox is deadly. As you've probably heard on the news, approximately one-third of the victims died and the survivors were scarred for life. This disease is spread by air droplet like the common cold. Antibiotics for virus diseases came along after smallpox was eradicated, so they haven't been tried against smallpox. Vaccination is available, but has not been routinely given after 1979. Recall from the section on vaccinations that this is a live-virus vaccine and does infect you. Also recall you're infected with cowpox from the vaccine.

Rabdovirus or Rabies virus

Rabies has several phases:

phase 1 with the symptoms: malaise, loss of appetite, fever, headache, avoidance of light, nausea, vomiting and sore throat.

phase 2 has the symptoms: nervousness, fearfulness, hallucinations, dilated pupils and tears, sweating, drooling and painful swallowing. People and animals infected with rabies avoid drinking because swallowing is painful, hence the other name for rabies—hydrophobia.

phase 3 has the symptoms of seizures, coma, and paralysis. Death results from respiratory paralysis.

Rabies is spread by animal bites. If a person is bitten by an animal that cannot be found and quarantined, or is suspected to have rabies, the person must be vaccinated immediately for rabies. Once the symptoms start, it is too late for treatment.

Rabies is a virus that can infect any warm-blooded animal. Many other viruses infect only one or a few species of animals.

Rotaviruses

Rotaviruses cause **diarrheal illnesses** with the symptoms: diarrhea, vomiting, fever, abdominal pain and dehydration from the vomiting and diarrhea. This may be fatal in infants and children. These infections need to be treated with fluid replacement and electrolytes, such as those from electrolyte drinks available at the pharmacy. Sports drinks may also be effective, but you should talk to your doctor about using sports drinks. This disease is spread by fecal contamination.

Rubeola-Measles virus

Rubeola-Measles virus causes **measles** with the symptoms: nasal congestion (coryza), cough, conjunctivitis, fever, bluish sores in your mouth that are called Koplik's spots and a rash that turns into crusted sores. This virus may cause encephalitis, pneumonia or ear infections. The disease is spread by air droplet. Antibiotic treatment is available. Vaccination is available and required for school age children in many states.

Slow viruses

Slow viruses cause **degeneration of brain**. The incubation period is years and sometimes decades. The degeneration leads to loss of muscle control and neurological problems, such as vision and hearing disturbances. The degeneration is eventually fatal. This disease is spread by from body fluids and tissue.

Tumor viruses

Some viruses, called tumor viruses, are known to cause **some human cancer**. Cervical and liver cancers are associated with viruses. Cervical cancer can be spread by sexual contact.

Prions

Prions are not complete viruses. They are the protein shell only. Prions have no genome.

Prions cause **spongiform encephalitis** with the symptoms: brain degeneration, loss of muscle control and strange behaviour–hence the name "mad cow" for the form that attacks cattle. There are forms of prion diseases that attack most mammals. This disease is spread by contact with blood or tissue of infected animals.

The human prion disease, Cruetzfeld-Jakob disease, has been spread by human growth hormone or tissue transplants.

There is no treatment or vaccination available for prion diseases. They are always fatal.

Chapter Ten
Parasite Diseases

The germs known as parasites are **protozoans** (one celled animals), **helminths** (worms) and **arthropods** (insects, such as fleas, ticks, lice). Most parasites can be found in the intestinal tract during part of their lives. The medical lab checks all fecal specimens for parasites, as well as for bacteria or fungi. Because the parasites are found in the intestinal tract, they are spread by fecal contamination. Some parasites may spread by other means, such as insect bites.

Unless noted otherwise, there are antibiotics to treat the parasitic infections.

Diseases caused by parasites include:

1. **Infectious jaundice** is caused by *Babesia microti.* It is spread by tick bites. The symptoms are nausea, fatigue, joint pain and depression.

2. **Diarrhea that alternates with constipation** is caused by *Bolantidium coli.* The disease is chronic and may go on for a long time. There may be blood or mucous in the stools.

3. ***Chilomastix mesnili*** is a sexually transmitted parasite. The symptoms are discharge and tenderness of infected area.

4. *Dientamoeba fragilis* is another intestinal parasite that causes **diarrhea**, vomiting weight loss and weakness. It is transmitted in water.

5. *Entamoeba hystolytica* causes an **intestinal disease** whose symptoms are abdominal cramping, loss of appetite, diarrhea, vomiting and nausea that result in weight loss. The disease may be mild or severe with dangerous dehydration. This infection can spread through the body. The parasite is found in feces.

6. **Giardiasis** is caused by *Giardia lamblia*. This is an intestinal disease that has the symptoms of diarrhea, malaise, weight loss, weakness, abdominal pain, gas and cramping. This is often a camper's disease, because the germ is found in water that has not been treated with chlorine to kill germs.

7. *Isospora belli* causes **intestinal disease** with symptoms of diarrhea, fever, abdominal pain and weakness. The diarrhea and fever may last for months.

8. *Leishmania donovani* attacks the body's protective cells, especially the macrophages in the spleen, liver, lymph nodes and bone marrow, causing **kala azar**. This disease leads to emaciation and weakness with intermittent fever. It is fatal if untreated. The parasite is spread by insect bites.

9. *Leishmania species* cause **sores at the site of insect bites**. The parasites spread to the nose and throat. The parasites may eat away the septum dividing the two sides of the nose, or the cartilage of the ears. The infections may be fatal because the throat and windpipe may become blocked by the damage done by the germ.

10. *Microsporidia species* cause **intestinal, eye and systemic infections**. These parasites are spread by fecal contamination.

11. *Plasmodium species* cause **malaria**. Malaria is transmitted to people from mosquito bites. The symptoms are chills and fever, nausea, vomiting and headache. The fever spikes then goes down for a while. The patient feels better while the fever is down, then the symptoms repeat. The symptoms may cycle through the fever and feeling better, many times. Malaria is treatable. The way to control malaria is to eliminate mosquitoes.

12. *Sarcocystis species* produce a toxin, called sarcocystin. The toxin causes **swelling under the skin and heart failure**. The parasite is found in undercooked meat.

13. *Trypanosoma* species cause **African sleeping sickness, chagas' disease and trypanosomiasis**. The parasites get into the body or the conjunctiva of

144

the eye from the person rubbing or scratching insect feces into the skin or eyes. The parasites cause fever, swollen eyelids and lymph nodes. When the parasites get into the blood, they spread through the body and attack the organs, or release toxins that damage the organs. The parasite can cause meningoencephalitis, which may be fatal. This parasite can be transmitted mother to child across the placenta. Some *Trypanosome* infections are treatable.

14. *Trichomonas vaginalis* causes **infections of the vaginal area and cervix** in women and **prostate infections** in men. The symptoms are discharge and tenderness. You might have guessed that this is one more sexually transmitted disease, and you'd be right. It can also be passed from mother to child during birth. Condoms provide protection against this disease.

15. There are amebas that are found in water that is contaminated with soil. These amebas can infect swimmers when water gets up their noses. The parasites spread to the **brain** causing **hemorrhages and damage**. The parasites can also infect sores and contact lenses and spread to the brain.

Helminths, better known as worms, **can attack any part of the body**. Different worms like different organs of your body. Helminth infections are treated with drugs or surgery to remove the worms.

The main source of helminths is fecal contamination of food or water. Proper cooking kills them in food. Treating water with chlorine kills them in the water.

Helminths include: heartworm, pin worm, tape worm, round worm, trichinosis, hook worm and flukes. In many parts of the country, pets should be given heartworm medication to prevent infection. Trichinosis is most often spread in undercooked pork.

Chapter Eleven
Fungus Diseases

You'll recall from the chapter on scientists and the germs they work with, that there are two types of fungi that fall into the category of germs. One type is **mold**. A common example of mold is the fuzzy, green stuff that grows on leftovers that have been in the fridge too long. The other type of fungus is **yeast**.

Fungi cause several types of diseases. Those types are:

Superficial mycoses that include:

1. Cutaneous mycoses–**infections of skin, hair and nails**

2. Subcutaneous mycoses–**infections below the skin**. These infections occur when fungi get into open wounds. Fungi growing under the skin or in wounds, cause growths that resemble warts or granular growths.

The superficial mycoses by name or site of infection are:

a **Athlete's foot** with symptoms of burning, itching feet.

b **Toenail fungus** whose symptoms are yellow or thickened nails.

c **Jock itch** which is itching in the groin area.

d **Ring worm** characterized by patchy red areas. When ring worm is on the scalp, it can cause bald spots.

e **Pityrinsis versicolor** whose symptom is the presence of patches of darker pigment on skin.

f **Seborrheic dermatitis** and **dandruff** that result in flaking of the skin or scalp are caused by *Malassezia furfur.*

Systemic mycoses

Systemic mycoses are often lung infections, though they can infect other organs. Molds are often the cause of the systemic mycoses. Bird and bat droppings carry the molds and yeasts that cause many systemic mycoses. You become infected by inhaling the mold spores or the yeasts.

These diseases are like flu in the mild forms and pneumonia when the infection is severe. Severe systemic infections have the symptoms of: swollen lymph nodes, enlarged liver and spleen, high fever, anemia, sores in your mouth, nose or intestinal tract, cough, joint pain, headaches. The infections can spread to other parts of the body, including the brain. They are more severe in immune compromised patients. They can be fatal if not treated with antifungal antibiotics.

The diseases and the fungi that cause them are:

a. **Coccidiomycosis**, caused by *Coccidiodes imitus*, a mold.

b. **Histplasmosis**, caused by *Histoplasma capsulatum.* Mortality is high when histoplasmosis is not treated with antifungal antibiotics. This is another mold.

c. **Blastomycosis**, caused by *Blastomyces dermatitis.* This is a yeast.

d. *Cryptococcus neoformans* causes a **pneumonia**. It is a yeast.

e. **Aspergillosis**, caused by mold *Aspergillus niger.* This mold also causes allergic reactions. It often grows on peanuts.

f. *Pneumocystis carnii.* It is another **pneumonia** causing fungus.

g. **Mucomycosis**, caused by *Mucorales* molds, starts as a sinus or nasal infection. As the infection develops, the mold filaments or hyphae invade all parts of the head and face, including the eyes and brain. The infected areas swell and may ooze bloody pus. Surgery as well as antibiotics may be required to treat this infection. Because of its ability to invade your body, this disease is often fatal.

Mucomycosis is more common in people who are diabetic, on dialysis or are suffering from debilitating diseases. Burn patients or people whose immune systems are compromised are also more vulnerable to this type of infection.

h. **Paracoccidiomycosis** is caused by *Paracoccidiodes brasiliensis*. It is a granuloma of the lungs, that may spread to skin, mucous membranes, lymph nodes, liver and spleen.

i. *Candida albicans* is a yeast that causes a variety of diseases. The disease that gets the most attention is **vaginal yeast infection**, whose symptoms are burning and itching of the vaginal area. This yeast also causes **thrush**, a discolored sore that usually forms on the tongue.

   Candida albicans also causes **skin and nail infections and systemic infections**. The systemic infections most often strike people whose immune systems are compromised. The systemic infections usually attack the skin, eyes, heart and membranes covering the brain, though they can attack any part of the body.

j. Some fungus infections are called **opportunistic mycoses**. These infections are caused by molds and yeast that are found in nature. These fungi usually don't cause disease. Like the bacteria called opportunist pathogens, these fungi cause infections only if they have the *opportunity* to do so. When your immune system isn't functioning at its best, you may be giving these fungi the opportunity to infect you.

Norcardiosis is an example of an opportunistic mycosis. It is usually a pneumonia, but can also mimic tuberculosis, with chest pain and weight loss. It may cause brain infections as well.

Toxins

Some fungi produce toxins. These are called mycotoxins. The mushrooms that are toxic, also produce mycotoxins. Aflatoxin is a mycotoxin produced by the mold, *Aspergillus niger*. Mycotoxins do not lose their toxic properties when they are heated, the way botulism toxin does.

Aflatoxin is one of the toxins suspected of being in Iraq's arsenal of biological weapons.

Allergies

Fungi may trigger allergic reactions. Even dead fungi can cause the allergic reactions.

Other problems caused by molds:

Molds growing in homes has been making the news quite a bit recently. Mold might cause a fungal disease of some kind or an allergic reaction.

Keep in mind that despite all the attention on the news, mold inside your home isn't the only place you're exposed to mold. You can get a fungal disease or have an allergic reaction from a mold that is outside your home. You'll find molds in the protected wetlands around some cities, in bird poop, compost, the garbage cans and just about any other place you can think of.

Fighting mold in your home:

You can fight back against mold in your home with a standard household disinfectant— **chlorine bleach**. You can dilute the bleach by using 1 cup of bleach to 9 cups of water to clean areas that can grow mold, such as around showers and shower curtains and bathtubs or sinks. Other household cleaners may kill mold. You'll have to read their labels carefully to find out.

You need to check for and repair leaks under your sinks or around windows to prevent mold from growing in those areas.

Mold can grow in wet laundry that is allowed to sit in the washer for a few days. You'll also find mold growing in wet bathing suits that you bagged up and forgot to un-bag when you got home from the beach or pool. Mold will also grow in your kitchen garbage.

Chapter Twelve
Epidemiology—Medical Detective Work

Epidemiology is the study of disease statistics. Sound boring? It's not. If you like mysteries, you'd like epidemiology.

Epidemiololgists do keep track of numbers. Those numbers are one of the tools they use to solve the medical "whodunits" or more correctly, "whatdunits."

There are two types of epidemiololgical studies. One type of study looks at long-term effects of something, such as smoking. These studies will look at the number of people who smoke and for how long, then compare that to the number of smokers who develop lung cancer or emphysema.

The other type of study looks at things in the short-term. This is the kind of study done when there's a flu outbreak. This kind of study was done when the recent anthrax letters appeared, as well.

Let's look at some example mysteries that an epidemiologist would solve.

Mystery #1. 500 people in a city become ill with hepatitis A. Hepatitis A is often transmitted in food. It has an incubation period ranging from approximately two weeks to two months, with an average incubation time of 1 month.

The epidemiologists who interview the people who got sick will want to know:

1. did anyone else in your family get sick?
2. What restaurants did you eat at during the past two months?
3. Did you travel to another city or country in the past two months?

Let's keep this one simple and assume that none of the victims left the city in the past two months. Let's also limit the restaurants in the city to: Carrie's Coffee Shop, Dean's Deli, Barney's Buffet, the Burger Place, the Taco Place, the Chicken Place, The Downtown Hotel Restaurant, the Pizza Delivery Place and The Organic Orgy.

Here are some example interview responses:

Person #1 gets coffee and a bagel every morning at Carrie's Coffee Shop, lunch at Dean's Deli every day except for the day 3 weeks ago when he took a prospective client to the Organic Orgy for lunch.

He took his family to Barney's Buffet for Sunday dinner 1 month ago and the Downtown Hotel Restaurant for brunch after church every other Sunday. No one else in his family is ill with hepatitis A.

Person #2 and his wife both have hepatitis A. They've had pizza from the Pizza Delivery Place 5 times in the past two months. They've eaten at Barney's Buffet twice and had fast food 10 times from the Burger and Taco Places. They celebrated their anniversary with dinner at the Organic Orgy six weeks ago.

Person #3 brown bags lunch every day. He gets breakfast sandwiches from the Burger Place every morning on the way to work. He has had dinner at home every night except when his brother visited. They ate at Barney's Buffet, Carrie's Coffee Shop and the Organic Orgy. His brother is also sick, but doesn't show up on the statistics for this city, because he lives in another city.

When all the interviews are done, the epidemiologists, start looking for a common restaurant that all the victims have patronized. What they find is that:

400 of the victims ate at Carrie's Coffee Shop

100 of the victims ate at Dean's Deli

350 of them ate at the Burger, Taco and Chicken places

115 of them had pizza delivered by the Pizza Delivery Place

92 of them ate at the Downtown Hotel Restaurant

397 of them ate at Barney's Buffet

500 of them ate at the Organic Orgy

The one restaurant that all the victims patronized, is the Organic Orgy. This is the common link.

The epidemiologists go to the Organic Orgy and test everyone working there for hepatitis A. They find that the two chefs and most of the other food handlers have hepatitis A.

If no one of the currently working at the Organic Orgy had hepatitis A, the epidemiologists would check former employees of the Organic Orgy for hepatitis A. If neither current nor former employees of the Organic Orgy had hepatitis A, the epidemiologists would have had to start looking for another source of the infection, such as the gourmet bottled water served by the Organic Orgy.

Mystery #2. The annual flu watch is on. A new strain of flu is working its way through Europe. Shortly after Thanksgiving, there are suddenly two separate outbreaks of this strain of flu in the U.S. and two victims. The first victim is in Laramie, Wyoming. She's the teenaged daughter of two University of Wyoming (UW) professors and a student at Laramie High School. The second victim is a first class cadet at West Point.

The epidemiologists go to work. Because the two outbreaks are so far apart geographically, the first assumption is that the virus got to the victims from two different sources.

UW has a fairly large population of foreign students. The epidemiologist guesses that the girl's parents may have carried the virus home to their daughter. He interviews the foreign students to see if anyone had gone home for the Thanksgiving break or had visitors from home. No one did. Next, the epidemiologist talks to the UW faculty, to find out if there were any visiting researchers or lecturers from outside the country. Again, the answer is no.

The other epidemiologist interviews the West Point cadet who had the dubious honor of being the first flu victim at that location. The cadet had gone home to New York City for Thanksgiving prior to getting sick. The epidemiologist checks to see if there are any reports of flu in New York City.

There are no *reported* cases of the flu at UW or in New York City. This doesn't necessarily mean there are no cases of the flu yet. It may mean that the victims just didn't go to their doctors. If you don't go to your doctor when you have the flu, you aren't included in the disease statistics.

The epidemiologist at West Point checks the records to find out if there were any foreign lecturers. Again, there are no likely flu carriers found.

The epidemiologist working in Laramie visits Laramie High School, where half the student body and faculty are now out with flu. He talks with the faculty members who are still on their feet. He asks about visitors to the school. There were no official visitors, but a graduate of Laramie High was in town to spend Thanksgiving with his parents. He stopped by the school to visit the day before Thanksgiving.

Further inquiry reveals that the Laramie High grad was also a West Point grad. He'd been in Europe doing research for his Master's Degree. The epidemiologist sends that information to CDC. CDC passes the information to the epidemiologist working at West Point. By now, West Point classes have been canceled for the week, because so many faculty members and cadets are sick with the flu. The epidemiologist interviews the faculty to find out if one of their grads came by to visit recently. A grad came to visit two days before the Thanksgiving break started.

The epidemiologist gets the name of the grad. It matches the name of the Laramie High School grad who had visited the school while home for the holidays. The flu's delivery boy has been identified.

Author's note: This is loosely based on the details surrounding a flu epidemic that simultaneously started at a high school in a western state and one of the U.S. military academies in the late 1970's. The details have been altered to protect the identity of the culprit.

Mystery #3. You be the epidemiologist. 1000 people are attending a convention. Meals are not provided, but morning and afternoon snacks are. There is also a "Happy Hour," where drinks and appetizers are served. 827 people become ill with symptoms of vomiting, diarrhea and fever by the 4th day of the convention.

Morning snacks consist of coffee, tea, juice, soda, bottled water and bagels. Afternoon snacks are the same drinks as the morning snacks and cookies or brownies.

The appetizers served during happy hour are:

chicken wings and ham sandwiches on Monday

ribs and chicken sandwiches on Tuesday

mini-tacos and roast beef sandwiches on Wednesday

egg rolls and meat balls on Thursday

boiled shrimp and cheese on Friday

Lab tests on the victims indicate *Salmonella* food poisoning. It has an incubation period of 24-48 hours and 100% morbidity, which means that everyone who is eats the contaminated food gets sick.

The illness starts on Thursday, so the days to focus on are Tuesday and Wednesday when you interview the victims, but you want to check everything for the week. The numbers you get are:

987 people ate the mini-tacos

1000 people ate the roast beef sandwiches

937 people ate chicken wings

827 people ate chicken sandwiches

990 people ate ham sandwiches

1000 people ate meat balls

1000 people ate shrimp and cheese

Which food was contaminated with *Salmonella*?

If you said chicken sandwiches, you're correct.

Mystery #4. The reunion picnic. 100 people are in attendance. Foods served are:

fried chicken

hot dogs

Darralyn McCall

potato salad

pasta salad

brownies

cookies

ice cream

72 people got food poisoning with the symptoms of vomiting and diarrhea. No one has a fever, which suggests *Staphyloccocal* food poisoning. This disease has a morbidity rate of approximately 85%, which means that a lucky 15% of the people who ate the contaminated food don't get sick.

Let's analyze how many people ate each food and how many of them got sick.

Chicken. 100 people ate it and 72 people got sick, 72%

Hot dogs. 97 people ate them and 72 got sick or 74%

Potato salad. 85 people ate it, 72 of them got sick, or 85%

Pasta salad. 88 people at it, 63 got sick, or 72%

Cookies. 98 people at them, 70 sick or 71%

Brownies. 66 ate them, 54 sick or 82%

Ice cream. 100 people ate it, 72 sick, or 72%

Which food caused the food poisoning?

Did you say potato salad? If so, you're right.

Conclusion

Knowledge is power. It is as effective against terrorism as it is against mundane diseases, like the common cold.

If you read the entire book, you now know more about the germs you've been hearing about in the news or have read about in the news magazines or in thriller novels. You'll be able to judge for yourself whether the author or journalist did his homework.

You have also learned where you can get more information about the germs and about using disinfectants. In case you didn't mark those pages, here are the URLs again:

http://www.cdc.gov

http://www.epa.gov

http://www.chlorox.com

Suggested Further reading:

For general information, see any textbook on Microbiology or Medical Microbiology that is available from an online bookstore or a college bookstore.

For the quick general review, see:

Introduction to Microbiology from the Blackwell Science, Inc. Eleventh Hour Series by Darralyn McCall, David Stock and Phillip Achey

To learn more about the history of microbiology or the impact of germs on history, see:

Viruses, Plagues and History by Michael B. A. Oldstone published by Oxford University Press

Killer Germs, Microbes and Diseases That Threaten Humanity by Barry E. Zimmerman and David J. Zimmerman published by Contemporary Books

Rats, Lice and History by Hans Zinsser published by Little Brown & Company

Flu by Gina Kolata published by Farrar Straus Giroux

To learn more about Preparedness against Bioterrorism, see:

When Every Moment Counts, by Senator Bill Frist, M.D. published by Rowman and Littlefield Publishers, Inc.

To learn more about the structure of germs and "cell activities," see:

Microbial Physiology and Metabolism by Daniel R. Caldwell published by Star Publishing Company

GLOSSARY

acidosis low levels of bicarbonate (carbon dioxide) in your blood. You can cause acidosis by breathing deeply and rapidly, a technique that is called hyperventilating. This is why some people have to breathe into a paper bag for a minute or so if they have panic attacks and breathe rapidly.

acute means severe and short-lived, as in a cold is an acute infection.

anaerobic living in an environment that has no oxygen. Some bacteria are anaerobic, such as the bacteria that live in a cow's rumen (second stomach).

anemia low hemoglobin in the red blood cells or a low number of red blood cells. This condition reduces your blood's ability to carry oxygen and it causes fatigue and weakness.

angiomatosis this is when blood vessels grow out of control into piles of "blood-vessel spaghetti," whose technical term is a "tumor-like mass." These piles of blood-vessel spaghetti can form in the skin, bones, liver or any other organ.

antibody a blood protein that is produced by the immune system. It is also called a gamma-globulin.

antigen a substance or part of a germ that causes your immune system to create antibodies. The protein shell of flu virus is an antigen, so is the venom that bees inject into you when they sting you.

antigenic type one species of germ, bacteria in particular, may have many variants that cause different kinds of antibodies to be formed. A good example of this is *Streptococcus pyogenes*. Group A (the flesh-eating bacterium that also causes strep throat), causes your immune system to make one kind of antibody. Group G, which causes strep throat, but doesn't eat your flesh, causes your immune system to create a totally different antibody.

aplastic anemia is anemia that is caused by destruction of your red blood cells or something that causes your body to stop making red blood cells.

arthalgia is joint pain.

aseptic meningitis is an infection of the membranes that cover your brain. It is usually caused by viruses, because there are no living germs found in the spinal fluid and no pus is produced. Bacteria are the germs that usually cause pus to be produced.

atrophy is wasting of part of your body.

bacteremia is when there are bacteria in your blood, but they aren't growing there. The germs are just riding along in your blood stream to get to some place to cause an infection. This is different from septicemia when bacteria are growing in your blood.

cellulitis is inflammation of cellular tissue.

cerebrospinal fluid (CSF) better known just as spinal fluid. This is the fluid that circulates around your brain and spinal cord.

choriomeningitis another term for aseptic meningitis. Meningitis is inflammation of the meninges—the membranes that cover the brain and spinal cord.

chorioretinitis inflammation or retina or membranes covering your eye and eyelid.

chronic diseases are those that last a long time. They are often less severe that acute infections.

colic is a sudden sharp pain somewhere in your abdomen. Sometimes colic is called a paroxysmal pain which is just the medical term for a sudden sharp pain.

colitis is inflammation of your colon or large intestine.

congenital conditions are those conditions that are present when you're born.

conjunctiva is the mucous membrane of your eye and eyelid

consolidation means that something becomes solid. It often refers to the pneumonia symptom when part of your lungs become filled with thick, gooey mucous and lose their stretchiness.

coryza is the cool-sounding medical term for a stuffy, runny nose, like you experience when you have a cold.

cystitis is inflammation of your bladder.

debilitated is the medical term for being in a weakened condition.

disseminated means spread out, like a rash that spreads over your entire body. Campaign propaganda that everyone has heard can also be termed disseminated.

distention is stretching. A fully inflated balloon is distended.

dura mater is the outer membrane that covers your brain. The dura mater looks more like bone than a membrane. The other two of the three membranes that cover your brain are delicate and "wispy." One of the wispy membranes is called the arachnoid mater, because it is like a spider web (scientific term for spider is arachnoid).

dysfunction means malfunction or impaired function.

dysuria is painful or difficult urination.

ecthyma gangrenosum is a mild skin infection that scabs over and may cause some scars.

edema is the medical term for fluid build up or "water retention." The fluid may build up in tissue spaces, body cavities or around joints and cause swelling.

embolism is blockage of blood flow. The blood flow block may be caused by a blood clot, fat glob, air or a piece of body tissue.

empyema is an accumulation of pus in some body cavity, such as your chest.

encephalopathy is brain disease.

endemic diseases are diseases that are always present in a small population or a geographic area. For example, bubonic plague is endemic in prairie dogs in Wyoming and Colorado.

endocarditis is inflammation of the membranes that line the chambers of your heart.

endometriosis problems with the tissue that lines the uterus.

epidemic refers to disease that hits many individuals at the same time. Flu is epidemic every winter, when millions of people are sick around the same time.

epidemiology -the short definition is that this is the study of disease statistics. Epidemiologists look at the causes and spread of diseases.

epithelial tissue is your skin and mucous membranes.

erythema nodusum is the medical term for a red lump on your skin.

erythemous is redness or inflammation.

exanthems is a disease that causes "eruptions," pustules or sores and a fever. Measles, chicken pox and smallpox are examples of examthems.

exudate pus or sweat—something that oozes from your body.

exudative is the fancy word for oozing.

fibrous or made of fibers.

filamentous made of strings or filaments.

follicular hypertrophy enlarged hair follicles.

fulminant diseases are those that suddenly become serious. Pulmonary anthrax is a fulminant disease.

gangrenous is tissue death or destruction. Something that interferes with blood circulation or causes pus and "rotting" leads to gangrene.

germinates begins to grow. Spores, like plant seeds, germinate.

gingivitis inflamed gums.

glomerulonephritis is kidney disease that attacks the capillaries that filter your blood.

granuloma is a tumor that is made up of grainy tissue or an inflammation.

hemorrhagic bloody or bleeding.

hepatitis is inflammation of your liver. Your skin may become yellow or jaundiced when you have hepatitis.

hydrophobia in reference to germs, is another name for rabies. If you're talking psychology, hydrophobia is fear of the water. Animals that have rabies avoid swallowing because it is so painful and because they avoid swallowing, they also avoid drinking water or any other liquid. This is how rabies got the name of hydrophobia.

hyperemia is an abnormally large amount of blood accumulating in some part of your body.

hyphae are the long threads of molds and *Actinomyetes.*

immunosuppressed is a term that means the immune system isn't fully functioning. Transplant patients are given drugs to cause immunosuppression to prevent transplant rejection. Rejection is when their immune systems attack the donated organs.

incubation is the period between the time you're infected by the germ and when you begin to experience the symptoms of the disease.

infarction is an area of dead or dying tissue. You may have heard heart attacks being referred to as "myocardial infarction." This means that a blood vessel is blocked and a part of your heart is deprived of blood, so that part of the heart muscle dies.

inflammation is redness, swelling and fever in an area of your body. Usually the inflamed area hurts when it's touched. The inflammation is the result of infection or injury.

inguinal lymphadenopathy is swelling of the lymph nodes in your groin area.

ischemic means the blood flow is reduced to a part of your body. The flow may be reduced because your blood vessels are constricted (being squeezed) or something is blocking them from inside, like a blood clot.

jaundice is a yellow coloration of your skin and the whites of your eyes because there are bile pigments in your blood. It usually indicates a liver ailment.

keratoconjunctivitis is inflammation of the cornea and conjunctiva of your eye.

lassitude weakness.

lesion a sore or area of diseased or abnormal tissue.

leukocytosis increase in the number of white blood cells.

leukopenia decrease in the number of white blood cells.

lymphadenopathy chronically swollen lymph nodes.

lymphocytosis increase in the number of lymphocytes—a type of white blood cell.

lymphoma a tumor of the lymph nodes.

macule a spot on your skin or an opaque spot on your cornea.

maculopapular inflamed bump on your skin. A zit is maculopapular in the early stages.

malaise that weird feeling you get when you're coming down with something.

meninges the membranes that cover your brain and spinal cord.

meningismus encephalitis without inflammation.

meningitis inflammation of the membranes surrounding your brain.

myalgia muscle pain.

necrosis tissue death. Usually this is localized around an injury.

neonatal newborn.

neuritis inflamed nerve. The symptoms of neuritis are usually pain, numbness or paralysis.

neuro a prefix meaning nerve or brain, as in a neurosurgeon being a doctor who operates on your brain and spinal cord.

nosocomial infections that you get in a hospital or doctor's office or from medical treatment.

ocular eye or if you're talking about a part of a microscope, the ocular is the part you look into.

opaque cloudy.

oropharyngeal mouth and throat.

osteitis inflammation of your bones.

osteomyelitis inflammation of your bones and bone marrow.

otalgia medical term for earache.

otitis inflammation of your ear. This is usually related to an earache.

pandemic this is an epidemic that is spread throughout the world. Flu is often pandemic, because people around the world get sick from it.

papule skin bump that is usually red and inflamed. Zits can be called papules.

parasthesia weird tingling or prickling sensation. When your feet go to "sleep," you're experiencing parasthesia. Parasthesia can be a symptom of disease or injury to nerves.

parotid gland the salivary gland that is below your jaw and ear. This is the gland that swells or becomes painful with mumps. Because this is the affected gland, mumps is also called parotitis—inflamed parotid gland.

paroxysmal means sudden attack.

perineum groin area.

periostitis inflamed periosteum or the covering of your bones.

petechial tiny hemorrhages in your skin or mucous membranes. Petechial hemorrhages look somewhat like a rash. They are a symptom of hemorrhagic fevers like Ebola.

pharyngitis sore throat.

photophobia sensitivity to light. Light may cause pain. This is a common migraine symptom as well as a symptom of some kinds of infections.

pleurodynia a medical term meaning that you have pain around your torso.

pneumonia inflammation and congestion in your lungs.

polyarthritis arthritis or inflammation of many different joints of your body.

prosthetic artificial or replacement body part.

pruritis itching.

pseudomembrane this is literally a false membrane. The disease diphtheria is characterized by the formation of a false membrane in your throat that is made from dead cells.

purpura reddish-brown or purple skin rash that is caused by petechial hemorrhages.

pustules small skin bumps that are full of pus.

radiculomyelopathy a problem with a spinal nerve that is often caused by irritation, rather than inflammation.

radiculopathy a nerve problem.

reticuloendothelial cells are protective cells that eat the invading germs.

rhinocerebral nose and brain.

saprophytic germs are those that live on dead stuff. The germs growing in compost are an example of saprophytic germs.

seborrheic dermatitis skin inflammation with heavy skin oil production.

secretions stuff your glands and cells release. Sweat, saliva and insulin are all secretions.

sepsis germs or their toxins attacking your body.

septicemia bacteria growing in your blood.

serotype a germ that is identified by the antibody it reacts with. This is another term for an antigenic type.

shock severe reduction of blood circulation. Shock can be caused by blood loss, infection or trauma. The symptoms are rapid pulse, pale skin, low blood pressure and cold clammy skin.

spongiform looks like a sponge that you use to wipe kitchen counters. This is a description of the brain of cows who are infected with "mad cow" disease.

spongiform encephalitis a type of encephalitis or brain infection, that causes the brain to take on a sponge appearance.

subcutaneous under the skin.

submandibular below the lower jaw.

synctial a mass of cytoplasm that has more than one nucleus.

systemic whole body.

tenesmus straining to urinate or defecate, usually without succeeding.

thrombocytopenia decrease in the number of blood platelets.

thrombosis blood clot in a blood vessel.

toxemia blood poisoning or toxins in your blood.

toxoid a non-toxic form of a toxin that is used as a vaccine. An example is tetanus toxoid used in the DPT shot.

transplacental crossing the placenta from mother to child or vice versa. AIDS can be spread transplacentally–mother to child.

trophic refers to how germs eat.

ulcer or ulcerative sore on the skin or mucous membrane. Tissue disintegrates and pus may be produced.

versicolor variant color.

vesicular or vesicle fluid filled sacs or cysts.

www.ingramcontent.com/pod-product-compliance
Lightning Source LLC
Chambersburg PA
CBHW032010170526
45157CB00002B/624